to Robin
Best regards

Fieldwork

Christopher Scholz

Fieldwork

A GEOLOGIST'S MEMOIR OF

THE KALAHARI

PRINCETON UNIVERSITY PRESS

PRINCETON, N.J.

Copyright © 1997 by Christopher Scholz
Published by Princeton University Press,
41 William Street, Princeton, New Jersey 08540
In the United Kingdom: Princeton University Press,
Chichester, West Sussex

Library of Congress Cataloging-in-Publication Data

Scholz, C. H. (Christopher H.)
Fieldwork : a geologist's memoir of the Kalahari /
Christopher Scholz
p. cm.
ISBN 0-691-01226-1 (cloth : alk. paper)
1. Scholz, C. H. (Christopher H.)—Journeys—Botswana—Okavango
River Delta Region.   2. Geology—Botswana—Okavango River Delta
Region.   3. Okavango River Delta Region (Botswana)—Description
and travel.   I. Title.
QE22.S34A3 1997
55.883—dc20   96-42052

This book has been composed in Palatino
Designed by Jan Lilly

Princeton University Press books are printed on
acid-free paper and meet the guidelines for permanence
and durability of the Committee on Production
Guidelines for Book Longevity of the
Council on Library Resources

Printed in the United States of America
by Princeton Academic Press

10 9 8 7 6 5 4 3 2 1

To Yoshiko

Nature, to be commanded, must be obeyed.

Francis Bacon,
Novum Organum
1620.

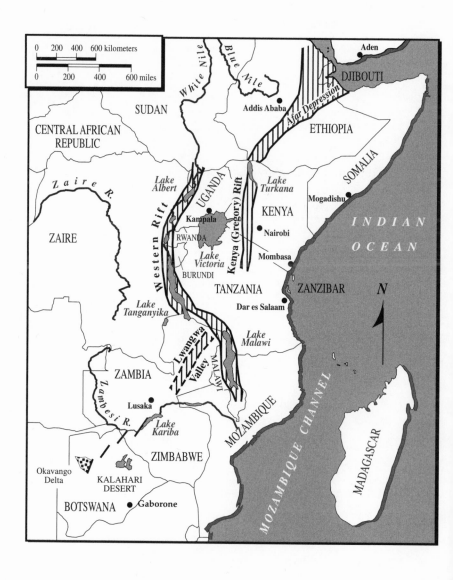

Fieldwork

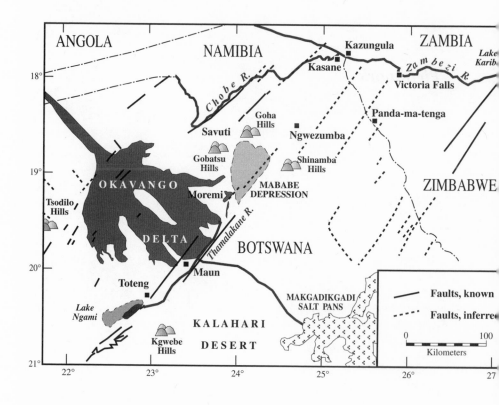

1

Teddy and I were sitting about twenty yards apart. We had been like that for more than an hour, hunched up against the trunks of a couple of mopani trees as we waited for the herd of elephants to leave the grove we were in. They had moved into the grove from several directions and by the time we had noticed them we had lost any chance of retreating back to the Land Rover. There was nothing for us to do in the meantime but wait them out and to try to be as inconspicuous as possible. Climbing a tree was no refuge in this situation. That offers protection from cape buffalo, but not from elephant, which can reach the upper branches of trees with their trunks.

The elephants were not showing any indications of leaving anytime soon. The trees were full of succulent fruit and the elephants were munching away on them contentedly, pulling whole branches down for their young. Even under the trees it was swelteringly hot, and we were being tortured by tse-tse flies, which were biting us incessantly. We couldn't swat them away for fear of attracting the attention of the elephants and alarming them. Teddy had managed to squirm around to the far side of his tree, so that we could look out in both directions. We weren't worried that we would be trapped there until dark. There was no water in the grove and sooner or later the herd would head off to be near a water hole by late afternoon. Our big fear was that one of the small grazing groups would select one of our trees and be startled by our presence. There wasn't much we could do in that case. You can't outrun an elephant. One time a mother elephant came close enough to catch our scent in the still air, and snorting, wheeled away, whisking her calf in front of her with her long trunk.

We didn't have any water with us, and thirst was starting to claw at my throat. With dry lips I gazed desirously at some half-rotten fruit lying on the ground ten feet away, but I didn't dare try to go get them. I simply concentrated on watching the elephants and breathing smoothly, stifling any tendency to sneeze or cough.

After about two hours, trumpeting rang out from the south. The elephants near us perked up their ears and listened a moment, then headed off in a brisk trot in the direction of the trumpeting. They had been summoned for their afternoon trek to water: "Lunch break's over. Everybody back on the bus."

Finding ourselves alone at last in the grove, Teddy and I stood up and looked around warily. When we had convinced ourselves that there were no stragglers left, we started hiking back together to where we had left the Land Rover beside the track. "One thing I can say about you, Scholz," said Teddy. "You sure can pick the places to go to study earthquakes."

2

It was in the austral spring of 1974, and I had my reasons for being out there trying to record earthquakes in the hot sun of northern Botswana. In any scientific endeavor there are several types of reasons for taking on one particular project rather than another, just as there are different sets of goals. There are the goals of the customer, the agency or company that is paying for the work, and there are also genuine scientific goals, in which one seeks to solve an original problem which will contribute to the understanding of some subject. These may or may not coincide; that depends on the customer. If they do not, as in this case, the trick is to devise a research strategy that will accomplish both for the same price and effort, which usually means solving the basic scientific problem that addresses the customer's problem as a particular case. Whether research is applied or basic is often a matter of point of view and manner of execution. They are not necessarily mutually exclusive.

In some cases, and these are to be recommended, there are deeper-seated reasons for doing a project. These are more personal and intangible than the others. They might be called personal development goals: You take on a project because of the feeling, perhaps inchoate, that it may in some way contribute to your deeper understanding of the larger-scale research program you have chosen as your life's work. This type of motivation is hard to assess and even harder to evaluate later on in terms of its influence on your scientific development. It is just something that is felt, and doesn't bear thinking about too much. Pasteur said, famously, "Chance favors the prepared mind," and this is one way the mind is prepared. This type of feeling was a prime motivation for my taking on the Botswana earthquake project. Some of these hunches pay off, often in unexpected ways, and some don't, but they all add the kind of seasoning to life that a wild card does to a poker deck.

Not all scientists have such a well-defined vocation—that scientific subject to which they have decided to devote a lifetime of work. Many, perhaps most, scientists are simply people who have mastered a certain set of specialized skills and apply these to whatever work in which they may be profitably employed. In my case, however, I found my vocation early and have managed to find ways to pursue it, with few diversions, ever since.

I study the mechanics of the brittle deformation of the Earth. The outer fifteen to twenty kilometers of the Earth is composed of rock that is cold enough that when it deforms, it does so by brittle fracture, forming the shear cracks that geologists call faults. When further deformation causes the walls of the faults to slide past one another, this frictional sliding is often unstable and occurs in jerks, producing earthquakes. These two related phenomena, earthquakes and faulting, are the central topics of my research. My approach to them is nontraditional. According to the traditional way the earth sciences are organized, these two topics are the subjects of disparate disciplines: seismology, which purports to be the study of earthquakes, and structural geology, which encompasses the study of faults. Thirty years ago, when I went off to graduate school at M.I.T., I chose not to concentrate in either of those traditional fields, but instead to pursue a Ph.D. in the more arcane field of rock mechanics, the study of the mechanical behavior of rock. I chose this approach because the body of knowledge and skills taught in the traditional fields of seismology and structural geology provide the training only to describe their subjects, earthquakes and faulting, and not to explain how they work. I was not interested in the "when" and "where" questions, but in the "how" and the "why." This latter class of questions could be answered only if you first understood the basic material properties that govern the mechanical behavior. Thus, to my way of looking at the problem, faulting was a form of brittle fracture, and earthquakes a form of frictional instability, and both of these properties, fracture and friction, depend on properties of the material: the rock. The correct approach, in my opinion, was to study rock mechanics first, the other disciplines later. That was my starting premise and the way I arranged to be trained. Developments since have served only to strengthen this conviction. Modern theories of earthquakes and faulting are now all grounded in rock mechanics principles.

I work on fundamental problems of earthquakes and faulting, which are universal, rather than regional, in their application. Most of my work tends to be experimental or theoretical rather than field-oriented. When I use field observations, it is to check or demonstrate theory. If I study an earthquake in Japan, for example, it is because there is something in the observations of that earthquake critical to some theoretical point, and not because I am intrinsically interested in Japanese earthquakes. Although I am known in my field as an expert on Japanese earthquakes, this is mainly because Japan is the scene of multitudes of diverse types of earthquakes that Japanese scientists have been carefully studying for centuries, producing one of the world's greatest treasuries of earthquake data and lore. I have spent years sifting through this treasure-house of data and have written a number of papers on Japanese earthquakes and tectonics, but it would be more accurate to say that I am a student of Japanese datasets than of Japanese earthquakes, per se. I am a net user rather than a generator of field data.

I seldom do fieldwork myself nowadays, but things were not always like that. More than twenty years ago, when the events I describe here took place, I went through a period of several years in which I did quite a lot of fieldwork. I was under the impression at the time that I needed to become acquainted with my subject first-hand, in the field, to enhance my intuition of the subject and to properly appreciate the field observations made by others. Whether or not this is true is another story. I have never regretted my years doing fieldwork. That kind of work entails encountering trials and challenges of a qualitatively different cast than are found in other kinds of scientific work, and good fieldworkers develop their own special set of traits and skills to cope with them.

At the time that I took on the Botswana field project I was a freshly minted and tenured thirty-year-old associate professor at Columbia University, working at the Lamont-Doherty Earth Observatory in Palisades, New York, known more familiarly as Lamont. My career was pretty well on track, but I wasn't yet senior enough to have gotten bogged down in administrative duties. Back then, in the early '70s, federal support for scientific research was still in its heyday. A more free and easy attitude prevailed in selecting research topics than today, when the funding squeeze keeps most academic scientists in a tight caucus race of grant getting and paper production that precludes forays into risky serendipitous pursuits. So the prevailing conditions made it easy for me to take on the Botswana project, oddball though it had seemed at the beginning. It would be a stretch to say that it was very closely related to my main line of research. It was not something I had sought out; it had really fallen into my lap from out of the blue. But as I proceeded to investigate it, my curiosity was aroused, and I found it had greater depth than it had first appeared. The problem the customer needed solving was, from a seismological point of view, pedestrian, but it had a potentially important impact on a vital environmental problem. The scientific problem involved was, on the face of it, solely of regional interest, but there was a chance, a slight chance, that this little region of northern Botswana held the key to a much grander problem that has puzzled geologists for more than a century: How do the continents rift?

The customer was the U.N. Food and Agricultural Organization (F.A.O.), an outfit with which I had had no previous dealings. I was introduced to it by way of a phone call from an official at the U.N. Headquarters in New York who simply asked me straight off if I was interested in serving them as an earthquake consultant in Botswana. It was not the sort of question I was prepared to answer immediately, if for no other reason than because at the time I had only the vaguest idea of where Botswana was. I asked for more information, trying to winkle this out in the conversation, but he said he knew nothing about it. He was just calling me as requested by a cable from the F.A.O. headquarters in Rome. I replied that I might be interested, and to please ask the F.A.O. to send me a description of the consulting assignment, along with whatever background information that went with it.

Botswana? No doubt that was somewhere in Africa. Where, I wasn't too sure. My formative education in geography was from a wall map that had hung in my room as a boy, which as far as Africa went was from a decidedly earlier era. I remembered vividly the configuration of all the former colonies, like the Anglo-Egyptian Sudan and French Equatorial Africa, even down to their colors. The former was a pale green, the latter a Rand-McNally orangey-red, which was why I had had the notion for a long time that the former was grasslands and forest and the latter desert. I had had no particular reason to study Africa later on, so I still hadn't quite caught up with the postcolonial names and changes in borders.

I was soon put right by the Times Atlas. Botswana was the old British Protectorate of Bechuanaland. Independent since 1966, it is located north of South Africa and between Namibia and the country that was then, in 1973, still called Rhodesia. The bulk of Botswana consists of the Kalahari Desert. This partly explained my ignorance. My geographical knowledge had become specialized to seismically active areas. I had great familiarity with such earthquake-prone regions as the Kuriles and Kamchatka, New Hebrides and the Kermadecs, but to my knowledge the Kalahari was not among the seismically active areas of the world. The only seismically active region in Africa was the East African Rift System, which was, or so I thought, located far to the north of Botswana. The wall map I lived by at the time, "Seismicity of the Earth," confirmed this impression: It didn't show any seismicity in southern Africa. So why did the U.N. need an earthquake specialist in Botswana?

A packet arrived from Rome the following week. In it was a letter inviting me to become a consultant on earthquakes for the U.N. Development Programme on the Okavango Delta. Attached was a beautiful full-color brochure that described, in general terms, the Okavango program. It was full of photographs of an incredible place that looked like a lush tropical version of pine barrens. They showed forests of trees on sandy islands surrounded by papyrus-choked channels, and herds of elephant and antelope grazing near pools of improbably clear water in which hippos wallowed.

According to my atlas, which referred to this place as the Great Okavango Swamps, this was a huge area of freshwater swamp that lay, enigmatically, in the middle of the northern Kalahari. The brochure informed me that these swamps contained the only natural fresh surface water in Botswana, and were therefore considered a key to the future development of the country, particularly with respect to its abundant but as yet unexploited mineral resources. Various schemes were evidently being proposed for tapping these waters. At the same time, the vast migratory wildlife herds of the Kalahari depended on the Okavango for their sustenance during the long dry periods between the brief rainy seasons. There were, obviously, serious environmental conflicts in this situation, so the F.A.O. had undertaken this development program which involved making a broad ecological study of the region with the aim of assessing the impact of removing any portion of the Okavango waters.

So what did all this have to do with earthquakes? I read the letter again. It stipulated that in serving as an "earthquake consultant" that I would be expected to spend some time at the program headquarters in Maun, Botswana, and to prepare a report at the conclusion of my work. There was no hint at what the earthquake problem was or what, for that matter, I was expected to do once I arrived at Maun. I cabled Rome for clarification, but received back only an uninformative message stating that the need for the services of an "earthquake consultant" was called for in the master plan for the Okavango project, and that was why I had been contacted. It implied that I could expect to be filled in on my duties once I reached Botswana.

This was very strange. Hiring consultants is expensive, so a consulting assignment usually comes with a very well-defined charge. Here there seemed to be no charge at all. It reminded me of the instructions Lord Copper, the newspaper magnate in Evelyn Waugh's *Scoop*, gave to the neophyte reporter Boot on sending him off to cover a rumored war in Abyssinia: "Let me see. You will get there in about three weeks. I should spend a day or two looking around and getting the background. Then a good, full length dispatch which we can feature with your name." I was getting the impression that the U.N. people themselves didn't know what their earthquake problem was. It was almost as if they were saying, "Hey, you're the earthquake expert, you figure it out!" This was my first encounter with the sort of bureaucratic vacuity that was typical of the organization I was soon to be dealing with.

There was no way I was going to go traipsing off to Botswana on such a poorly defined mission, but my curiosity had been aroused. So I decided to do my own investigation of the tectonics in the region. I went downstairs to the computer center and loaded the tapes containing the locations of all sizable earthquakes that had occurred since 1962, when the Worldwide Standardized Seismographic Network had been installed. I instructed the computer to plot out a map of earthquakes in Africa. In those days plotters were mechanical devices in which pens, moving on mechanical arms, drew the plots on paper. After first making an outline of the continent, the plotter pen whizzed back and forth, pausing here and there and with a thwat-thwat making a cross locating the occurrence of an earthquake, the size of the cross indicating its magnitude. As the computer ran through its data file, the pen spent most of the time plotting crosses in the northeast part of the map, indicating earthquakes that had occurred along the East African Rift System.

With a length a sixth the circumference of the Earth, the East African Rift System is in the process of splitting the African continent into two parts. It is most prominent in the north, in Ethiopia, where it is opening the fastest. There the block containing Somalia is splitting away from the rest of Africa in an east-west direction. This split is marked by a series of deep rift valleys, containing many active volcanoes, that runs south from the Red Sea along the center of the broadly uplifted region that forms the Ethiopian highlands. Just south of Lake Turkana in northern Kenya the rift divides into two branches, an eastern one continuing to the south, passing to the west of Nairobi, and a western branch that follows the edge of the Congo Basin, forming the valleys filled by the great African lakes: Albert, Kivu, Edward, and Tanganyika. East of Lake Tanganyika the two branches rejoin and continue south through the rift valley occupied by Lake Malawi and the Shire Valley of Malawi, which are the last recognizable traces of active rifting.

The little crosses that were appearing on the map were accurately delineating this tectonically active feature, which is what I had expected to see. Occasionally, though, with a whirr the pen would transit far to the south and west and leave its telltale mark. So southern Africa was not as aseismic as I had thought. I made another map, this time just of the region south of Lake Tanganyika. On this map a faint pattern could be seen. There was, to the east, an irregular pattern of crosses extending from Tanzania through Malawi and petering out in northern Mozambique. This marked the southernmost and least active part of the rift system. To my surprise, I could also make out another, fainter line of epicenters farther to the west. This trend that I was now seeing was something new. Tearing off the computer plot, I overlaid it on a map of Africa and saw that this trend ran in a southwesterly direction from Lake Tanganyika, following the course of the Lwangwa Valley in northern Zambia, then continued along the course of the Zambezi gorges on the Rhodesia-Zambia border. There it was highlighted by a dense cloud of epicenters in the Kariba Gorge of the middle Zambezi. Following this trend with my eye, I found a few scattered events farther to the southwest, ending with a small cluster of four epicenters in the vicinity of the Okavango Delta. Bingo! Something indeed was going on down there. I was starting to feel the little tingle I get whenever I have spotted something interesting that nobody has noticed before.

I went into the archives where we keep our library of original paper seismograms and files of earthquake catalogs from the pre-computer era. Digging through the old catalogues, I found that two magnitude 6.5 earthquakes had been located in the Okavango in the early '50s. Unfortunately, very little was known about them. The coverage and quality of seismographic stations in those days would not permit a modern seismological study of such remote events, and, hardly surprisingly for that time and place, no contemporary field investigations had been mounted. Then I remembered that in the early '50s Lamont had installed a small global network of the first matched set of modern long-period seismometers. This had been the predecessor of a larger network set up for the International Geophysical Year in 1957–58, and which later evolved into the Worldwide Standardized Seismographic Network. In the part of the library that housed the records and data from that era, I found a map of this network showing that it had included three stations in Africa: Johannesburg, South Africa; Fort Lamy, Chad; and Helwan, Egypt. Entering the stacks, where in floor-to-ceiling racks thousands of 3-by-1-foot flat cardboard boxes of seismograms were filed according to station name and date, I was able to find the original paper seismograms for these stations. I pulled the boxes for June–August 1954, when the earthquakes had occurred. I spread out the seismograms for the appropriate days on the analyst's table, and found to my delight that both earthquakes had been clearly recorded at each of the stations. By measuring off the amplitude and arrival times of the various seismic waves at these stations, I was able to recalculate the locations and magnitudes of the earthquakes, using modern computer methods. These recalculations confirmed the figures given in the old catalogue, showing that the Okavango region was indeed capable of producing quite large earthquakes. The question was: Why?

I had noticed that the Okavango seismicity came at the end of a weak trend of seismic activity extending all the way from the East African Rift System east of Lake Tanganyika. Could this indicate a previously unrecognized branch of the rift system? The structure of the Lwangwa valley certainly resembled a poorly formed rift, but the continuity of this trend all the way to Botswana might be an illusion. The dense cluster of seismic activity in the Kariba Gorge I knew to be a man-made artifact. A high dam had been erected in the gorge during the '60s, and most of this seismic activity had been induced by the filling of its reservoir. The impoundment of large reservoirs has often been observed to induce seismic activity below them, as a result of the combined effects of the weight of the impounded water and of its hydraulic pressure in changing the state of stress underground. The induced activity at Kariba had been the subject of a number of detailed investigations using a local seismic network that had been specially installed for the purpose, and there was no doubt about the nature of those earthquakes. But now, armed with a new viewpoint, I could ask a different question of the data obtained by this local network. Was the reservoir-induced seismicity the result of triggering of small earthquakes in an otherwise inactive region, or was it an enhancement of activity in a naturally active belt? Going back to the archives, I found the monthly catalogues of the Rhodesian Seismic Network, which had been set up to study the Kariba seismicity. Close examination of the data given in those catalogues, with this question in mind, produced a definitive answer. I found that small earthquakes had been located in a trend along the gorge that extended as far as fifty to sixty kilometers *downstream* of the dam. The effects of the filling of the reservoir on earthquake activity would be negligible this far away from the reservoir, so these had to be natural earthquakes. There was then no doubt in my mind that there was, along the Zambezi gorges, a zone of natural seismic activity that simply had been stirred up by the filling of Lake Kariba.

The clincher for me arrived the following week, in the form of Landsat images that I had ordered from NASA. It is always spectacular to see these satellite images. Each one covers a region of about forty thousand square kilometers (200 x 200 km) and is printed in false colors diagnostic of moisture, vegetation, and rock and soil type. They show a level of detail greater than any map and over a scale range broader than can be visualized by any other means. To a geologist they are nothing short of wonderful: a treasure trove of information. They are called images, rather than photographs, because they are made using a number of spectral bands, several of which are not in the visual range, and because, having been gathered by several types of electronic instruments and constructed and reproduced by computer, they have little in common with what is produced with ordinary cameras or film.

The images were made at a time of the year, common for this region, in which there was no interrupting cloud cover. They showed an amazingly flat region, devoid of almost any topographic feature except the occasional blemish of a kopje, the isolated little rocky hills of southern Africa that poke out of the desert like islands in a sea of sand. Along the northern edge of the region the Zambezi flowed from west to east, but had not yet entered, at Victoria Falls, its series of deep gorges, and so flowed, desultorily, in a meandering series of swamps and braided streams. In the northwest corner of Botswana, in the area bordering Angola and the narrow salient of Namibia called the Caprivi Strip, there was a vast empty region covered by long linear sand dunes running east and west. Through this meandered, in a generally southeastern direction, the Okavango River, which drains the well-watered Angola highlands. Then suddenly, in this almost totally flat terrain, the river split into hundreds of channels, which diverging, formed an ever expanding filigree of anastomosing streams and pools. It looked as though an enormous glass of water had been spilled over the desert, and every point that had been touched by the water had turned bright green.

More astonishing still, a hundred miles or so downstream from this point of divergence, all of these myriad streams again rejoined, collecting into a river perpendicular to the first, and forming this whole watery region into a roughly equilateral triangle, some 120 miles on a side. This was the Okavango Delta. I could also see why this was a freshwater delta. It had two outlets, one to the south, which terminated in the saline Lake Ngami, and one that continued farther to the southeast, finally ending in the great Makgadikgadi Salt Pans.

What a curious situation this was. For very good reasons, rivers do not generally branch in a downstream direction. And, under the combined effects of gravity and erosion, they usually flow directly to the lowest elevation in the region, where, if there is no outlet, they evaporate to form a salt flat or saline lake. They do not usually form inland freshwater deltas like this, perched above their natural hydrological declivities. And even if they do, they have one, and not two, outlets. This was not a stable situation. There was some other agent, beside the normal action of flowing water, involved in producing and maintaining this peculiar arrangement.

I pored over all the images. I had taken the precaution of ordering a full set of images at each individual spectral band, and in several combinations, since the different spectral frequencies are sensitive to different kinds of features. I first focused on the Thamalakane River, which formed the downstream base of the triangle of the Okavango, because it was suspiciously straight along much of its course. Looking carefully, I could also see subtle linear features running from southwest to northeast across a number of the images. Two of these ran from Lake Ngami across the base of the delta, between which flowed the Thamalakane, the river that collected all the delta streams. Parallel lineaments, offset in an echelon manner, continued through the veld of the Mababe and Chobe regions to the northeast as far as the border with Rhodesia and Zambia. Another, similar set of features could be seen at the head of the delta. These were all parallel to the trend of the line of seismic activity that passed through the Kariba Gorge and continued on to the Okavango. Bingo, and bingo again!

There are very few features in nature that are really straight, and at this scale, tens to hundreds of kilometers, there is only one likely suspect: a geological fault. If these lineaments were active tectonic faults, extending as it seemed in a trend from the basin of Lake Ngami all the way to the Zambezi gorges, then the most likely explanation was that I was indeed looking at the tip of a previously unrecognized branch of the African rift system. That would explain the seismic activity, and more. If these faults had the same sense of motion as the faults that bound the African rifts, then the Okavango Delta was located within a nascent, now barely perceptible, downdropped rift valley. This one was about a hundred miles wide, somewhat wider than the great rift valleys of East Africa; unlike those, the bounding faults in this one must have just started to move. Here there were no great escarpments dropping the valley floor a kilometer or so below the level of the surrounding plateaus. Whatever depression had occurred by fault motion was masked by being infilled with sand. At the head of the delta the river crossed several faults, which, I guessed, produced a sudden increase in slope in the downstream direction, contrary to the normal hydro-graphic profile of rivers, so the stream split there into many branches. At the distal end of the delta the waters encountered another fault which was slipping the other direction, with the north-west side moving down. That created a low scarp facing upstream, which acted as a natural dam, channeling the Thamalakane and em-pounding the Okavango water above the local minima in elevation: Lake Ngami and the salt pans. To maintain this arrangement these faults had to be actively moving to counter the normal erosional ac-tion of the rivers, which, if given enough time, would otherwise de-stroy these unstable features.

The hypothesis had the delightful feature of explaining things on two very different scales. Regionally, it could account for the pres-ence of that faint band of seismicity as delineating a previously un-recognized arm of the African rifts that extended all the way from Tanzania to somewhere just south of the Okavango. On the local scale, it also explained why the Okavango existed. This latter point tickled my sense of whimsy. Seismologists have known for some time about reservoir-induced earthquakes, as at Kariba, but here was the first recognized case of an earthquake-induced reservoir!

I was now beginning to realize that there was an intelligence be-hind this assignment, after all. Any earthquake occurring on one of these faults would produce fault motions that were mainly vertical. An earthquake as large as the ones of the '50s could be expected to produce a few meters of uplift and subsidence, which in such a flat area would be enough to seriously alter the drainage pattern within the Okavango. The whole hydrological system would be put out of plumb. Any development scheme that would seek to tap these wa-ters would have to accept the chance that it could be seriously dis-rupted by an earthquake of that size. Because the whole region was tectonically active, the topography was unstable and the stream pattern was therefore intrinsically ephemeral. The customer did have a potentially major earthquake problem, after all. What it needed to know was how likely this possiblity was, and how seri-ous its consequences.

I was later to find out that the person behind my assignment had been Colin Reeves, a British geophysicist, who, fresh out of the uni-versity, had done a brief tour of duty with the Botswana Geological Survey. At that time the Okavango program was still in its planning stages and had been passed around various Botswana government agencies for review. It was Reeves who had suggested that earthquakes be put on the menu for study. It was this recommendation that, after winding its way through some bureaucratic labyrinth and in the process losing its original rationale, had finally fallen on my plate, as though out of a rabbit hole. I also found out later that he had reached more or less the same conclusions as I about the origin of this seismic activity, but I was glad to have been left to work it out for myself, because now I had the bit well in my mouth. I was now mentally committed to go to Botswana. To prove the existence of a new branch of the African rifts was a nice little scientific problem. I had formulated a good enough working hypothesis, but to prove it I had to get more data, and to do that, I had to go to Botswana. What I had in mind, though, had little to do with what the F.A.O. had proposed in its letter. Simply to go to Botswana and spend a few weeks sitting around the U.N. Development Programme office made no more sense than Lord Copper's instructions. I wouldn't learn any more that way than what I already knew. The hypothesis I had was fine as far as it went, but what it needed was a little more data to flesh it out. And getting those data, under these circumstances, was not going to be a trivial undertaking.

I need to be clear about this. I had developed a hypothesis, which I liked very much. It had a certain beauty: It was consistent with the way I think the Earth works, it agreed with all the data that were available, and, as I mentioned, it offered explanations for things, otherwise seemingly unconnected, at both large and small scales. But it was still a very weak hypothesis. I was jumping to conclusions. I thought the lineations that I saw on the satellite images were active faults, but maybe they were linear sand dunes, like the ones I had seen in the northwestern part of the country. If I went to Botswana and just had a look at them, I probably would not be able to tell one way or the other. I had also guessed which way the faults were moving, this being at the heart of my "nascent rift valley" idea. This was partly by analogy to the rifts of northeast Africa and partly by a guess in interpreting the flow patterns of the rivers. But you can't prove anything by analogy, and there might well be another, more mundane explanation for the behavior of the rivers. Just because my explanation was grand, or "elegant," as scientists say, did not mean it was right. The way to test this hypothesis was with seismicity. If the faults were active, then earthquakes must be occurring on them, so that if I went to Botswana with sensitive seismographic equipment I should be able to detect and record this activity. If I could do that, I should be able to prove that the faults were active and also to determine the direction in which they were slipping, the critical test of my rift valley idea. This kind of fieldwork would have to be done, in any case, to properly assess the seismic risk in the Okavango for the U.N. program, but I would also want the project to be designed broadly enough to encompass the larger scientific problem as well, otherwise the job would hold no interest for me. That would require a lot more work than what the F.A.O. had in mind.

It was time to have a chat with Teddy. Teddy Koczynski, my technician, was the man who made sure that my various experimental and fieldwork schemes actually worked. He had been with me almost from my start at Lamont. After arriving fresh out of a post-doc and struggling for a year on my own, I had realized that I needed someone with better practical talents than I if I were ever going to build up a functioning and competitive Rock Mechanics Laboratory where I could do the kind of experiments I needed to understand the inner workings of earthquakes. At the time, Teddy had been working as a seagoing tech on one of Lamont's oceanographic research vessels, but his wife had put the kibosh on his spending six months a year at sea, so he was looking for a shore job. He had heard that I was looking for somebody, and turned up at my lab one day, a skinny little guy with a wild tuft of brown hair, scraggy mustache, wire-rimmed glasses, and an infectious grin. It was good timing for both of us, so I hired him on the spot, which turned out to be one of the best moves I've ever made. He was rated as an electronics technician, having learned that trade in the Navy, but a boyhood of tearing up and reassembling motorcycles and cars in a small town in south Jersey had equipped him with a pretty good set of mechanical and problem-solving skills. Whatever knowledge of hydraulics and high-pressure technology he needed, I taught him. I found him to be a quick study, and someone who knew that there is no time when you stop learning. This is essential for the kind of work we do, because it never gets routine. The Rock Mechanics Lab was built to do experiments deforming rock at the high pressures and temperatures that exist deep within the Earth where earthquakes occur. There are very few labs of this type, so the equipment in them is specialized and has to be custom built, which is far from a cut-and-dried job. Within two years he had mastered this technology and was running the lab as his own domain. My role had become that of dreaming up the experimental program and finding the money to pay for it, which suited me just fine.

I found him sitting at a bench, in a Sherwin-Williams painter's cap, poking at a circuit board with a soldering iron. "Hey, Chris, what's happening?" he said in the absent way he has whenever he is deeply absorbed in anything.

"Not too much. What would you think about a little field trip to Africa?"

"Africa? Heh-heh. Why not? Whereabouts?" I had gotten his attention.

"Botswana."

"Botswana? Hey, that's all desert down there, isn't it? The Kalahari. Bushmen and all that stuff. I saw a program on it on TV a while back. Pretty good stuff. But I didn't know there were any earthquakes down that way."

"I didn't either, until recently. The question is, why?"

"Ah-ha, another case for the seismic sleuths, eh? By the way, that reminds me, did you hear the one about the airline pilot who got lost and was running out of gas over the Kalahari?" I grunted. "So he sent out a mayday on the radio. The airport control tower answered it, saying, `Bechuanaland here.' The pilot answered, `You're damned right, but where the hell am I?' "

I made the required groaning noise. "Christ, I hope that's your only Bechuanaland joke. Come on, lets go look at some maps."

What I had in mind was something called a microearthquake survey. This technique takes advantage of one of the few empirical laws of seismology, called the Gutenberg-Richter relation. According to this law, in any seismically active region, there are always many more small earthquakes than large ones, following a very regular pattern. If the earthquake magnitude is decreased by one unit, there will be 10 times as many of the smaller- than the larger-magnitude events. So, for example, if a given region experienced, on average, one magnitude 6 earthquake every 10 years, you could expect, on average, 10 magnitude 5 and 100 magnitude 4 events in the same period. The little earthquakes carry the same information about the tectonics as the big ones, so the idea of a microearthquake survey is to install a local network of very sensitive seismic stations that are capable of recording microearthquakes, events in the magnitude 1 to 2 range, so that in a few months in the field you might register as many events as you would get in 40 or 50 years with the much less sensitive global seismic network.

The strategy Teddy and I concocted over the next few weeks was fairly standard. We would use about six portable microearthquake instruments and set them up in a local network fifty kilometers or so in diameter and record microearthquakes for a week or two, and then pick them up and move them to the adjacent area until we had covered the whole of the study area. Except for a couple of special instruments we had for very remote sites, the microearthquake units had to be tended once a day, so six was about the limit two people would be able to handle in that kind of country. The study area, from Lake Ngami to the Zambezi at the Zambia-Rhodesia border, was nearly five hundred kilometers in extent, but we figured we could do the whole thing in about three months if we pushed it hard and if there were no major hitches. This latter was a pretty big if, of course. It ignored one of the great rules of human endeavor and one that is particularly relevant to fieldwork: Murphy's Law, "If something can go wrong, it will." We were now proving its corrolary: "When planning any endeavor, you will underestimate the power of Murphy's Law."

Teddy and I pored over every available map of the region. By a little subterfuge, we had even gotten some semi-classified Air Force ones. They were all dismally bad, and they weren't very encouraging. This was a pretty wild part of the world. Most of it was in wild animal country uninhabited by humans. It was obvious that for most of the time we would have to live and work in the bush, often in remote areas accessible only by tracks in dubious condition. The logistics were going to be difficult, to say the least, and would require substantial support, which we had no idea how to arrange.

Nonetheless, I thought it had a fair enough chance of success, so I contacted the F.A.O. and said I would accept the assignment. My terms, however, would be different from those described in the letter. Instead, I proposed to carry out a microearthquake survey of three months' duration, and told them I was prepared to submit to them a fully costed out research proposal to that effect. In my letter

I explained why such fieldwork was necessary; still, this evidently caused much consternation in Rome, because they did not reply for a long time. They were simply not used to doing business this way. They were used to hiring consultants, but not letting out research contracts to universities, so when we finally reentered negotiations I had to explain how the university contracts and grants system works in the U.S. But in the end, after many confused communications and long delays, they told me go ahead and prepare a formal proposal.

In the meantime, I had been working on the nuts and bolts of the logistics: transport and field support, but with little help from the other end. I had written at some length spelling out our needs to the U.N.D.P. office in Maun and to the Botswana Geological Survey, with little in the way of satisfactory responses. Dr. Ernest, head of the program's Maun office, seemed unable to comprehend our plans, and kept responding to my increasingly persistent and detailed queries by reassuring me that we would be provided with a bungalow, office, and car while we were in Maun. I could not seem to be able to get him to understand that we were going to be in the bush for three months, and not sitting around an office in Maun, and that the support requirements for that were much greater and of an entirely different character.

I had better luck with the Botswana Geological Survey in Lobatse. After an initially taciturn response I was eventually put in touch with Dave Hutchins, Reeves's successor, who assured me that they would provide our transport and field support, although he wouldn't, even under prodding, tell me exactly what that would amount to. This was the most worrisome aspect of the whole project for me. Without adequate field support, we might well arrive in the field and simply not be able to carry out the work. It wasn't until we got to Lobatse that Dave told me why I had been getting such lukewarm and evasive responses. The people in the Botswana Geological Survey didn't want to commit themselves because they didn't believe that our project would ever get off the ground. Their basic attitude to our project, a popular topic around the Geological Survey tea room at the time, was something like the following: "The U.N. has got some crazy seismologist from New York who imagines he is going to do a microearthquake survey in the Okavango and the Chobe. Bloody daft idea, it can't be done! But there's no point in telling him that, he's some muck-a-muck professor at Columbia. Just wait till he comes down and he'll straighten himself out shortly. Haw-haw."

In my innocence I blithely wrote in my budget that the Botswana Geological Survey would pay for transport and field support and we would charge the U.N. only for our personal expenses and a modest amount for "contingencies." After more long delays, the F.A.O. finally approved my proposal, and the project was on.

3

Africa is a continent like no other. It is unusually flat and high, with vast reaches of its interior consisting of wide plateaus. One may find there escarpments that resemble serrated mountain ridges, but these are capped, not by a summit but by a broad highland. There are isolated volcanic edifices, like Mount Kilimanjaro, or the remote and forboding Ahaggar of southern Algeria, but there are no true mountain ranges, with the exception of the Atlas of Morocco and the Cape Fold Belt in the extreme south.

In the north, Africa is impinging upon Eurasia; but with the exception of the Atlas, the deformation produced by that continental collision is confined to the Alpine belt on the far side of the Mediterranean. On all other sides, the oceans surrounding Africa contain mid-ocean ridges along which the ocean basins are spreading, causing all the other continents to be moving away from Africa. This current plate tectonic configuration was completed in the Cretaceous period, some 65 million years ago, with the final split-up of the supercontinent of Pangaea, of which Africa was the ancient and protected core. Africa is *in toto* the oldest of the continents, most of it being formed in the Archean era, during the earliest period of Earth's history, more than 2.5 billion years ago. The Earth was hotter and more chemically primitive then, and the ancient cores of the continents, the Precambrian shields, were formed by massive upwellings of magma that solidified to form stable and resistant cratons of granitic rock surrounded like grout by greenstone belts where most of the deformation was concentrated in the subsequent jostling among the cratons. This is the structure of most of Africa. Tectonic episodes continued to deform the greenstone belts of Africa up through the Paleozoic era, but the effects of these events have not survived as mountain ranges, and they never produced a splitting of the ancient continent.

The exception to this long-lived stability is the system of great rift valleys of Africa, which, in the words of the American geologist Bailey Willis, "lie like inverted mountain ranges sunk in the plateaus." This system of continental rift valleys is the most spectacular and well developed of any to be seen on Earth today. These features, in which the central valley floor has been depressed, often by several kilometers, by slippage along steeply inclined bounding faults, have long been the subject of a great geological controversy. The first continental rift to be studied by the emerging science of geology was that part of the valley of the Rhine called the Rhine graben (from the German for ditch or grave), where it flows between the massifs of the Vosges and the Black Forest. Elie de Beaumont, the most prominent of the early French geologists, proposed in 1846 that the Vosges and Black Forest were once part of the same plateau that had been arched up, with the Rhine graben formed as a dropped keystone of the broken arch. This was contested a generation later by the influential Viennese geologist Edouard Suess, who was adamantly opposed to the idea of absolute uplift in geology, and who argued that the Rhine graben was formed by tension, in which the crust was extended, with no prior arching. The Scottish geologist J. W. Gregory, who pioneered the study of the African rifts and whose name is associated with the eastern rift in Kenya, was an ardent follower of Suess and interpreted the African rifts in the same manner.

These old theories, like patriarchs, have continued to influence younger generations of geologists. The grand theory of plate tectonics, which has encompassed and unified all aspects of tectonics, has placed the phenomenon of rifting in a new context but has not yet succeeded in solving the root problem of the cause and mechanism of rifting. Continents surely rift, and if the crust within the rift is thinned enough by extension, magma may be injected, forming the basaltic crust of the oceans, and true sea-floor spreading then begins. But what initiates this process, and what causes rifts to occur where they do, are among the large unanswered questions in geology. Many attempts have been made to answer these questions by studying rifts in different stages of development. The edges of the old rifts along which the Atlantic opened are buried beneath a thick drapery of sediments on the continental slopes that hide them from view, but younger rifts, like that of the Red Sea, which has recently split Arabia off from Africa, are more amenable to study. One can trace the Red Sea rift from the southeast, where true sea-floor spreading is occurring, into the Gulf of Suez, and via the Gulf of Aqaba to the Dead Sea rift, wherein neither the Gulf of Suez nor the Dead Sea has the continent yet to be fully sundered. There are also failed rifts, which became inactive without splitting the continent, and are left on land where they may be mapped and studied. The Rhine graben is one of these, as are the Newark Basin and the Connecticut Valley, which are part of a system of rift valleys bordering the east coast of North America that were active in the Triassic period, but were abandoned when a series of rifts farther east succeeded in opening the North Atlantic and splitting North America off from North Africa and Europe.

Uplifted mountain blocks border all of these rifts, but it is not clear the uplift was essential in causing the rifting or whether it was something that accompanied the rifting or was even a consequence of it. Nonetheless, the two main classes of post-plate tectonic theories, like those of the nineteenth century, focus on the presence or absence of doming or arching preceding rifting; though embellished by modern knowledge and ideas, the theories still echo the old controversy of de Beaumont and Suess. In one class of models, rifting is initiated at "hotspots," where plumes of magma rising from the deep mantle impinge upon the crust, causing uplift and magmatism. When such hotspots occur beneath oceanic plates moving rapidly over the mantle, they may leave behind trails of oceanic islands, like the Hawaiian chain in the Pacific. If they occur beneath rapidly moving continents, they may leave a wake of old volcanic activity, like that in Idaho's Snake River Plains, which marks the track of North America moving over the hotspot that is now beneath Yellowstone. Africa is almost stationary with respect to the mantle, so that the hotspot thought to exist beneath the Afar triangle in northern Ethiopia, at the triple junction of the East African rifts and the rifts opening the Red Sea and the Gulf of Aden, has remained in the same place since at least the beginning of the Miocene epoch, about 24 million years ago. It is this hotspot that is thought to have caused the doming up of the Ethiopian highlands and initiated the African rifts. But rifting must require an additional ingredient, the presence of crustal tension, because not all hotspots produce rifting.

If the East African rifts have propagated as far south as Botswana, that is too far from the Afar hotspot for the rifting to have been influenced by the forces associated with the hotspot itself. In other words, the initiation of rifting may have been localized by the hotspot and its associated doming, but the rifts continued to propagate southward under the action of crustal tension alone. This implies that rifting may not require prior doming, but may occur solely in response to crustal extension.

Plate tectonics is a kinematic model, meaning that it describes the motions of the plates, but not how or why such things as rifting occur. To understand such processes requires dynamic models, ones that include forces as well as movements. Problems of this class are more difficult to solve than the kinematic ones, and are the type that interest me. I was intrigued by the possibility that rifting extended into the Kalahari, because there one might be able to view the very tip of a propagating rift in an otherwise undisturbed setting. If my hypothesis was right, this spot was almost unique in the world. To study this place firsthand was my private agenda for taking on the Botswana field project. I was not so naive as to think that doing a microearthquake project in this region would enable me to solve the problem of rifting, but I did think that it would provide me with a unique perspective should I be in a position to tackle the main problem, later on in my career.

Once I had submitted my proposal into the black hole of the U.N.'s bureaucratic machinery, I had plenty of time to study the subject at leisure. Every year I run a semester-long seminar on some topic in tectonics, so in the spring of '74 I organized one on rifting. This is a good way of combining teaching with researching the literature on a subject of current interest. That semester half a dozen graduate students and I read through the entire modern literature on rifting, meeting once a week to critically discuss the week's readings. The discussion sessions were often also attended by several members of the faculty or research staff. Those people were often world experts on the weekly topic and their insider's knowledge added greatly to our appreciation of what points were then well established and which were still in debate, giving us a real flavor of the current state of knowledge. By the end of the semester, not only were the students well educated, but I was well up to speed on a topic that I had been scarcely acquainted with six months earlier. I was now intellectually prepared to go to the field.

It was almost a year from the fall of '73 when I had submitted my proposal when our contract authorization finally arrived from the F.A.O. By then it was the beginning of September 1974, and we were in a frantic rush to get started The rainy season in the Kalahari could begin as early as November, and that would turn the desert into a quagmire and put an early end to our field season. Teddy and I had already put in a month of feverish activity getting all the field gear collected and refurbished, since most of it had just come back from another project in Italy and was much the worse for wear. As soon as the grants office told us that the funds had arrived, we scrambled to get everything packed and air-freighted to Gaborone, the capitol of Botswana, where, following by a different route, we planned to pick it up. Of course in our rush we had forgotten to get inoculations, so we had to get the whole tropical suite the day before our departure. We got shots for yellow fever, typhoid, bubonic plague, and probably a few other things that I've since forgotten. We had so many shots that the nurse ran out of big fatty muscles to inject into and had to double them up, so it was two very sore and stiff fellows who boarded the plane for Rome the next evening.

The F.A.O. had insisted on issuing us our tickets, which turned out to be the nonrefundable kind, which itself told me something about the organization. What was really peculiar about the tickets, though, was their routing. They read: New York-Rome-Johannesburg-Gabarone-Selibi/Phikwe-Maun. The F.A.O. had explained to us the reason for the Rome leg; they wanted us to go there before leaving for the field, to be briefed at their headquarters. But when we tried to book the rest of the itinerary, we found that Gaborone was the only possible air destination in Botswana. Selibi/Phikwe did not yet have an airport, and Maun had no scheduled service. How they had managed to write these tickets and calculate the fares I'll never know, but I suppose that is why they wanted to issue them.

We had cabled Rome to let them know of our arrival and they had booked us into a small hotel convenient to the F.A.O. headquarters, which occupies a large modern building in the center of the city. The hotel was an old-fashioned place in a quiet neighborhood. Our room was spacious with large windows but not much light, owing to a queer arrangement of angles and shadows. It had an enormous marble-lined bathroom in the center of which was installed an elegant and complexly plumbed bidet. Though we knew in theory what this was, we had to try all its knobs to see how it worked, in the course of which we succeeded in douching down half the room. We were, after all, just a pair of typical American rubes on the continent. In the evening we had a memorable walk around the old neighborhoods in the Roman dusk, finishing up with a huge bowl of spaghetti a la carbonara in a sidewalk trattoria.

When first thing the next morning we reported to the big office block housing the F.A.O. headquarters, we had some difficulty in locating the right office for our briefing. A bemused receptionist sent us off to some office where we were directed on to another, and so on, until after a sequence of four or five such iterations we finally found someone who was aware of the U.N.D.P. Okavango Project. Far from expecting us, he didn't know who we were or why we had arrived in Rome. He told us that there were two people in his office who handled southern African projects, but they were both on vacation and there was no briefing scheduled for us. He further informed us that they didn't do briefings anymore, and that we should have gone directly to Botswana. But he insisted that as long as we were there we should pick up travel authorizations so that we could claim reembursement for travel expenses while we were in the field. This seemed totally unnecessary to me, since we had already included those expenses in our grant budget, but my trying to explain that met with blank looks of incomprehension, so we spent another hour or so waiting for the proper papers to be filled out.

I was furious by the time we left. By some stroke of bureaucratic idiocy we had been sent on a wild goose chase to Rome at a time when we were in a mad rush to get into the field. I fumed over the lost time, but we were in Rome and the weather was gorgeous, so it was hard to stay in a bad mood for long. We had a day and a half to kill before our flight to southern Africa, so Teddy and I set off to see the sights and enjoy a few last days of civilization.

There were two minor episodes in Rome worth recounting because, other than further examples of American rubishness, in retrospect they seemed in some ways to presage later events. That afternoon, while sipping aperitifs at a sidewalk café, and generally hanging out and trying to look cool, we were befriended by a group of five Roman teenagers. They sat down at our table and we had a lively conversation for a few minutes. Just after they had ciao-ed us and left, I glanced down and saw with a shock the empty space on the pavement where my camera had been. I leapt up and bolted down the sidewalk after them. Sprinting, I reached them just as they were climbing into their car, and leaping into the rear and sprawling across the laps of the three guys sitting there, I grabbed the middle one by the collar. In the next instant, Teddy had hauled the front door open and was threatening to put a beer bottle through the windshield. The boys, all bigger than we were, must have been astonished by our daffy display of aggression, because the camera was quickly forthcoming from beneath the jacket of the driver.

I was as elated by the outcome as I was embarrassed at having gotten into that idiotic situation in the first place. That incident said something about our naivete, but it also said something about our sense of teamwork. We had an instinctive congruence in thinking that resulted in a coordination of action of a sort that can never be rehearsed. This would stand us in good stead in the months to come.

The second incident, on the other hand, said more about the difference in the way Teddy and I were mentally preparing ourselves for our coming adventures. While strolling along the Via Veneto that evening, Teddy suddenly stopped, as though dumbstruck, in front of a sidewalk vendor who was demonstrating one of those plastic electric yo-yos that light up in garish flashing colors when spun. "Hey, I've got to get me one of those," he said. I was astonished. "Why in the world would you want one of those things? They're just tourist junk." He looked at me with a big grin. "Wait till we get down to Botswana," he said. "Boy, will I be able to show the witch doctors a thing or two with this thing, heh, heh."

We got on the flight to Johannesburg the next afternoon. It was a South African Airways flight that went via Lisbon and the Cape Verdes, because SAA wasn't allowed to fly over black Africa. At fifteen hours' flying time on a course almost due south, it was the longest flight I had ever taken without a time change. After flying through the night over the South Atlantic, we crossed the coast in southern Angola and cut across Namibia and Botswana on the final approach to Johannesburg. Just at daybreak we caught a glimpse, in the distance off the port wing, of the sun glinting off the waters of the Okavango. Then, as the sun rose, we flew over the vast brown desolate plain of the Kalahari. It was a breathtaking sight, but a bit daunting too, since we were soon going to be living and working down in that empty wilderness. We glanced at each other a bit uneasily, and didn't say a word.

I didn't much care for Jan Smuts Airport. It was my first encounter with apartheid and there was an almost palpable odor of it in the air there. Burly, submachine gun–toting police were everywhere. Security police frisked passengers in small cubicles, while big German shepherds sniffed their luggage. Crowds of shabby blacks squatted with their bundles in dark corridors away from the main departure hall where there were only white passengers and black porters. Toilets were marked Schwartzes and Blancks instead of Men and Women. You didn't need to have any preconceived notions of life in the Republic of South Africa to get a sense of the repression and conflict going on there.

At the Air Botswana counter they claimed they didn't know us. They said they hadn't received our reservations and told us that the flight that day was full. They furthermore informed us that they wouldn't be able to get us on one of their twice-weekly flights for at least three more weeks. What a pain! Now we would have to rent a car and drive to Gaborone, and after flying all night we were in no mood for ten or twelve hours of driving across the parched veld on half-paved roads.

Since we were tired and angry and a little hung over from too many drinks on the overnight flight, we decided to lodge a formal complaint with the airline before leaving. In the next building we found the office of the air service company that represents Air Botswana in South Africa. After hearing our story, the man in charge said, "Ja, so they said they didn't get your reservations, hey? Listen, those boys don't pay any attention to reservations. They just let anybody on who they think is the most important fella or has the most loose cash. I'll have a word with them, then." He dialed the phone and spoke quickly for a few moments in loud and brusque Afrikaans. Hanging up, he turned to us and said, "Right, you're on, then, hey? Just get over there quickly and get your bags checked before they claim the flight is overweight."

So we hauled our gear back to the main terminal and joined the colorfully bedecked mob that was camped, with all their various packs, sacks, and odd pieces of luggage, at the Air Botswana gate. Thanks to the intercession of the Afrikaner, we had no difficulty in checking our bags and getting onto that day's flight. The plane was a fright, an old Hawker-Siddeley of early '50s vintage. The baggage compartment in the nose was quickly filled, and so the plane was packed, not only with passengers but with what looked like their entire worldly belongings. As we levered our way into our seats, Ted remarked, "Jeez, no wonder they were worried about weight. I hope this crate can get off the ground."

We flew for two hours over the sun-baked brown plain, with just a few odd kopjes braking the monotony of the landscape. Descending, we got our first glimpse of Gaborone. It consisted of a few blocks of modern two- or three-story buildings surrounded by what looked like a field of brown beehives. As we got closer, these turned out to be rondavels, circular mud and wattle huts with thatched roofs, enclosed in mud-walled compounds. "Sure looks like bongo-bongo-land, all right," said Ted. "Suburbs," said I.

We taxied to the terminal, around which were parked a few small planes and a single charter jet sporting a flashy paint job. The terminal building was a tin-roofed shed of concrete block, open on the apron side. We joined the packed queue in the baggage and customs enclosure, which was separated by a chain-link fence from the other side, equally packed with a crowd clamoring to board the return flight. When we finally disengaged ourselves from whatever customs process that went on, we found ourselves standing, amidst all our gear, at the end of a dirt parking lot. Our fellow passengers milled past us, some getting into cars and some just walking off across the desert with their parcels. No one from the U.N. or the Botswana Geological Survey appeared to greet us. There were no taxis or buses. The return flight departed, and so did the airline and customs people. Soon we were standing in front of an empty terminal, facing an empty parking lot. We could see the town, a mile or two away across the desert. With a sinking feeling, we looked at all our stuff, trying to decide how we could tote it all that far. Just then, a pick-up pulled up, and the ruddy-faced driver yelled out at us,

"Hey, you boys look like you need a lift. Hop in." Gratefully, we piled our stuff in the back and climbed into the cab. "Where will it be then?" he said, as we headed off down the dirt road into town. "I suppose you boys will be wanting the Holiday Inn?"

What an introduction to the exotic dark continent! There it was, sitting out there alone in the desert with its hideous Holiday Inn sign planted in a tiny patch of regulation green grass and towering over a neatly whitewashed formulaic Holiday Inn building. Appearances can be deceiving, though, because inside it was anything but typical of the chain of American "family" hotels. The large lobby was dominated by a luxurious bar and casino. The air-conditioned rooms were served by closed-circuit TV that featured a selection of first-run films and pornography. The poolside was packed during the day, and the casino by night, where dusky beauties, clad in colorful and imaginative costumes, à la Frederick's of Hollywood, consorted at the tables with the players. The clientele consisted mainly of white South African men brought in by charter for sinful holidays just outside the borders of the repressive old fatherland. Tsk-tsk. Shame on you, Holiday Inns!

The next day was spent on various odds and ends in Gaborone. At the U.N.D.P. Office we had once again to explain ourselves, since in spite of our telegrams no one there was expecting us or knew of our project. The office chief, a tall Swede in an elegant blue suit, gave us a canned and uninspired little speech on the Okavango program and seemed at a loss at what to do with us until I presented our travel documents from Rome, which gave him a sense of direction. These papers seemed to serve as our letters of introduction with the U.N., and they prompted the preparation of more documents, but since the office was in disorder owing to the recent departure of some key personnel, we were asked to call back later to pick them up. Meanwhile we did a round of errands in the town, picking up small supplies at the hardware store and making what banking arrangements we needed. The great success of the morning was Teddy's. He managed to find a Chinese grocer who supplied him with stocks of soy sauce and garlic, prime ingredients for his secret recipe for steak teriyaki which he had been promising to cook for me once we got into the field.

We met Dave Hutchins in the bar at the Holiday Inn that afternoon. We watched him come in and look all around the nearly empty bar before hesitantly approaching us. No doubt he was expecting to meet a graybeard, whereas I wasn't more than three of four years his senior and didn't have the typecast appearance of a professor. He was a stocky Dorsetshireman with a full blond beard and ruddy face, on his first job out of university. We got along well immediately, even though I wasn't pleased when he told us that our "camp" wouldn't be ready for another week. In the meantime, he suggested that we accompany him back to Lobatse the next morning, where we could get settled in and have a look at whatever information and gear the Geological Survey had to offer that might be useful to our project.

Botswana is a country the size of France, and at that time it had a population of about half a million, and eleven miles of paved road. The road to Lobatse wasn't one of those. There had been record rains earlier that year that had flooded the desert, and every time the road crossed a calcrete flat it was filled with deep ruts that testified to the miseries of past travelers. As we rocketed along in Dave's Land Rover, he warned us about getting caught in those ruts at high speed. By chance, I had already heard the same warning. A year or so before, a friend of mine, Joe Hamilton, had been on a trip collecting rock samples for petrological study in the hills near Lobatse. While driving on that road back to Gaborone with a load of samples, he had caught a wheel in a sand-covered rut at seventy miles an hour. His van flipped and rolled across the desert floor, with him inside, revolving with hundreds of loose rock samples like a human in a ball-mill. Luckily he was not seriously injured, but what a mess of bruises he had collected!

Thinking about that, I looked around at the countryside with great interest. The basement rock of southern Africa is from the Archaen Era, dating from the earliest period of Earth's history. The rocks that Joe had been sampling were the oldest ever found, 3.8 billion years, and their chemistry contains evidence of the early evolution of the Earth's crust and mantle at a time when the Earth was much hotter and geologic processes very different than they are now. Even the look of the country had an alien, strange feel to it. We were traveling down a flat valley, bounded on each side by knobby ancient ridges of dark Precambrian rock that were so resistant to erosion that they ended abruptly at the valley floor, with only a few rounded boulders for scree. They were the remnants of a time when the Earth was a hot, roiling mass, just beginning to sort its primordial matter into crust, mantle, and core, the three great divisions of its interior. I had never seen anything that looked like this before. It was as though we were on a different planet.

As we bounced along, talking with Dave about the geology, and about other matters that interest newcomers, we were in excellent spirits. We were finally beginning to get that great feeling of freedom and release that comes with being in the field. Here was a life immeasurably different from that of the lab, office, and home. We had left far behind our ordinary daily routines, our familial chores and responsibilities, our humdrum worries. And yet we were not on a petty lark, a boondoggle, or a vacation. We did not have that feeling of organized release from ordinary life that comes with tourism. We had real work to do, and real responsibilities to go with it. But how different those were from those of everyday life. Instead of spending a day tracking down a bug in a computer program, fussing with the purchasing department about some equipment specifications and delivery times, or worrying about the kids' problems in school, for the next few months we would have to face problems of an entirely different type, like how to jury-rig a fuel line to nurse a broken-down Land Rover back to camp through a hundred miles of wilderness. People actually pay to go to wilderness camps, river trips, and so forth, just to get a taste for that sort of thing, even though it is all under safe and contrived conditions. Here we would be doing it for real, with live ammunition, so to speak. We were entering, for a time, a different kind of existence, one that we yearned for. After all, why does one go into geology in the first place, if not for times like these? We would be working in Nature, attempting to wrest from her some of her secrets, and prepared to contend with any obstacles she might place in our way. We were going on a safari, but not as fat-cat tourists hiring an expensive outfitter with nursemaid treatment in the bush and all the pleasures of a three-star hotel in camp. We were going to run our own safari, but we were after knowledge instead of animals. In a place like this, that would be a venture combining the most exalted and the most basic of human challenges. We had to solve a scientific problem, and, not to put too fine a point on it, we also had to survive.

4

The Cumberland Hotel, a massive pile in blistered and peeling pink stucco, squatted on a flat dusty expanse on the outskirts of Lobatse. As though to spite its unprepossessive form and setting, inside it presented a fading pretentiousness that seemed a relic of some glorious though ersatz colonial past. It had a grand dining room, august and empty, off a spacious foyer that managed to disguise the tawdriness of the guest accommodations, a row of shabby rooms tucked along a corrugated-roofed balcony overlooking a scraggy garden in the back of the building. The rooms were fitted out with cute amenities such as little guest sewing kits in matchbooklike containers inscribed with the name of the hotel, but were crammed with cheesy furniture and equipped with old-fashioned tank-type air-conditioners that were so noisy and drippy that one almost preferred to endure the sweltering heat than to turn them on.

These architectural details seemed a perfect reflection of the character of the Cumberland's proprietor, like the building itself a fine example of the triumph of form over substance. He was an expatriate Englishman with a sallow complexion and sourer disposition who affected the genteel airs of an earlier era. He was a hostelier in the "Fawlty Towers" mold. When we first arrived, he glanced distainfully at our rucksacks, duffel, and dusty clothes, and made a big fuss about our lack of reservations. He pretended to make a big effort to fit us in, giving the practiced impression that he would deign to let us stay in his establishment even though we clearly weren't the sort of gentlemen who were his usual clientele. This was all pretense, of course. We soon found out that there were only two other paying guests in the hotel, neither of whom looked gentlemanly enough to make Teddy or me feel out of place.

The centrum of the Cumberland's pomposity was its dining room. Fronting an entire wall of the lobby behind tall windows draped spinsterishly in chintz, it was a large room that occupied the whole two-story height of the building. Its exterior wall was also fitted with floor-to-ceiling windows that were mercifully covered in dingy velour drapes that hid from view the dusty and trash-bedecked tract of land behind the hotel. At the dinner hour, the black waiters, whom we recognized from their daytime roles as porters and swampers, could be seen, all decked out in scruffy satin striped pants, red jackets, and tassled red fezzes, standing about feigning attentiveness amongst the tables. Glass-topped dessert caddies and carts with chafing dishes were parked about, on the ready in case there was call for the serving of gala confections or flambéed dishes. A beat-up grand piano proudly occupied a small stage in one corner, as though ready to be joined at any time by a cotillion orchestra. But by far the most noticeable feature of the dining room was that it was almost always deserted.

On our first evening in Lobatse, Teddy and I entered this grande salle with the hesitation one has when entering any conspicuously empty eating establishment. "Why is no one else here?" is the sort of warning that rings in the mind at such times. It is best to attend to such inner trepidations. On the first occasion we didn't have the opportunity to find out, since we were denied entry on the grounds of improper attire. Jackets and ties were required for dinner, and although Teddy was able to find a ratty old tie in his "kit," I didn't have one with me. Hotel guests or not, we were simply not up to standards, or so the proprietor stiffly informed us. I later bought a grotesquely ugly tie from an Indian merchant in the town just so, for a change of menu, we could try out the dining room. What a treat! We sat through six courses of flamboyantly bad food made interminable by our waiters' slow and fussy impression of what they imagined was four-star hotel service, all of which would have been hysterically funny if we had not been its squirming victims.

Although it never would have been publicly admitted by the proprietor, who never set foot in the place, all the money the Cumberland made was in the bar—or rather, bars. There were three of them. The Lounge Bar, a relic of the days when ladies were obliged to drink separately from men, was a room inside the hotel that was seldom used except by the occasional hotel guest who wished to take a drink after hours. The main action was in the Private Bar, which was private only in the sense that British public schools are public. Its name served to differentiate it from the Public Bar, a dark dirt-floored shed along one side of the building where native sorghum beer and cut-rate alcohol were offered to a stand-up-only crowd. The Private Bar, in contrast, was the social center for the "Europeans" of the town, and those "locals" well enough up in the civil service to be associating with the Europeans. It had some of the atmosphere of a British pub, although the ratio of consumption of gins-and-tonics to beers was more in the tropical tradition. It at least offered a good pub supper, which provided Teddy and me a haven from the main dining room.

Lobatse is Botswana's southernmost railhead on the Rhodesian Railroad. It lies at the crossing of the railroad with the highway linking Gaborone and Mafeking, South Africa, which had been the administrative center of the old protectorate. In those days, before the diamond mines had come in, meat was Botswana's largest export, and the Botswana Meat Company's abattoir at Lobatse was its largest outlet. Every year a nationwide cattle drive took place, with the Lobatse stockyards as its final destination.

The business district consisted of a few blocks of Indian- and Chinese-run shops opposite the railroad. These were all-purpose emporia in big tin-roofed sheds behind railed sidewalks that stood three feet above street level, relics of the days of horses and cattle drives. The bulk of the town consisted of native quarters, irregular clusters of compounds of mud huts that gradually diminished in density with distance. Near the Cumberland at the north end of town were neighborhoods of government-issue bungalows of several grades, assigned according to the rank of the occupant. These, located on dirt roads extending for several hundred yards on either side of the main road, housed the whites and middle-class blacks. The most memorable feature of the town was the all-pervading odor of the stockyards.

The Geological Survey occupied a large compound of buildings near the center of town. On our first day, Dave introduced us around the place. The field manager confirmed Dave's bit of bad news about our "camp." It would be at least a week before it was assembled. Both the crew and vehicles for our project had to be pulled wherever available from other projects all over the country, since the Geological Survey was at the height of its field season and was already stretched to the limit of its logistics. On the other hand, he assured us that we would be supplied with the best of crews. The leader of our crew was to be a man named Blackie, who was the best and most experienced man they had. Dave commented on our luck on getting this fellow, who seemed to be something of a legend at the survey. He also gave us the reassuring news that our campkeeper, Deacon, didn't drink and wasn't too bad a cook.

Great. It seemed like we were in a hurry-up-and-wait situation. Our camp wasn't ready, and our gear, which had been air-freighted from New York a week earlier, had not yet arrived, either. It seemed to have disappeared somewhere into Lostluggageland. It was beginning to sound as if we would be cooling our heels for awhile in Lobatse. Everybody said, "Don't worry, everything will be ready in a week or two." A week or two! I wanted to be in the field by tomorrow at the latest! This wasn't the first or last time this sort of thing has happened to me. It's a matter of perspective. We had been rushing madly for the past month in a frenzy to get into the field to start collecting data. Now that we had arrived at the staging area we somehow expected everyone there to get up to New York speed and hustle to get things going for us. Not too surprisingly, people responded with attitudes like, "Hey, relax, man. What's the hurry? Take it easy." What is one man's field area is another's home base.

Pulling into the Cumberland parking lot on our second day in Lobatse, Dave was loudly greeted by some men sitting on the verandah, who beckoned us over. Seven or eight men were seated at a big round table littered with beer cans and half-consumed plates of food. Introductions were dominated by a great beefy florid-faced man with a bushy Boer mustache and loud voice to match. "Davey," he roared, "sit your arse down and squat vid us awhile, boyo. These your Yank friends, hey? Velcoum, velcoum, sit yourself down and haff a beer vid us, hey. Jake Horsfeldt, that's me. I'm the vet for the meat company, see, and I've given out 400 shots for hoof-and-mouth this morning, and planning on having a few shots of me own now, har-har."

He yelled at the waiter, "Another round here, Bobby, and for these boys, too. Celebration today, boys. Big sendoff for young Derek here, off tomorrow to protect the Fatherland from the bad buggers, eh, Derek?"

Derek was another sort of South African. Well-tanned and slim, in white tropicals and knee socks, he spoke in the clipped and measured accents of a British South African. He was a geophysicist for Anglo-American, the big South African mining conglomerate, running a uranium exploration field crew in the southern Kalahari. We got into a long discussion on fieldwork conditions and tactics in the desert. He was mainly doing aeromagnetic and gravity surveying, but he had done enough seismic profiling to give me the bad news: The sands were as deadening to seismic waves as a featherbed. "But if you can find a solid calcrete bed," he told me, "you've got a bushman's chance to get signal."

Jake's announced excuse for this "piss-up," as he called it, was at least true: Derek was indeed leaving the next day for the obligatory six weeks' annual service with the South African Army. At that time, there were wars going on all around southern Africa. There were internecine wars in both of the former Portuguese colonies, Angola and Mozambique, as well as a more conventional anticolonial guerrilla war against the breakaway white regime in Rhodesia. All these nasty conflicts tended to spill over in various ways into the neighboring countries.

"So what will you be doing, then?" I asked. It hadn't been all that long since our own miserable experience in Vietnam, so I was very interested in the attitudes of a South African conscript in defending the land of apartheid.

His attitude was phlegmatic and direct, not at all what I might have expected. "Border patrol, Mozambique, I expect. You get to traipse a platoon back and forth along 50 kilometers of border. It's mainly bloody boring, but you have to keep a constant eye out for the freshly planted mine. Strange that, being bored and watchful at the same time. It's just like being an inspector at a canning plant. Can't let any of the canned mice get through, or its curtains. Once in a while though, the RENAMO buggers, who have more or less control in the south, get funny and lob a few rockets over. That livens things up a bit. Nice for a change of pace. Unless somebody buys it, of course."

He had a remarkably clearheaded view of the politics of those wars, I thought. "The Portuguese just packed up and scarpered, leaving no one in particular in charge. So the various bully-boys are having a go at it and this will keep up until one of 'em kills the others off and takes the lot. God knows how long that will take. You Yanks, the Russkies, the Chinese, the Cubans, and who knows who else have all taken sides and make sure that their boys don't run short of the deadlies. But for most of the population in those countries, it's keep your head down and try to get fed."

"The one that interests us is Rhodesia. We all figure that Ian Smith's lot is going to lose out in the long run. The real question is, what happens afterward? When Mugabe's boys take over, as they're bound to sooner or later, the question is whether or not the whites will be able to carry on. The answer to that will say a lot about the future for us in South Africa."

The party, presided over by Jake and a few cronies, and refreshed from time to time by new arrivals, lasted late into the night, ending up in the Lounge Bar. Teddy and I finally dragged ourselves off to our rooms, fortunately not far away, leaving half a dozen revelers sitting around a bottle-laden table. That wasn't the only such boozy occasion we were to experience during our time in Lobatse, because drinking was a major activity among a fair segment of the white men living there. This usually began with a few drinks over a pub lunch, and then recommenced with sundowners in the afternoon, which, when I come to think of it, always seemed to precede that namesake time by quite a few hours. Teddy and I were generally invited for drinks several times a day, often by Jake, who was the most regular of the regulars at the bar. At first we thought this was a sign of great hospitality to us newcomers, but later we began to suspect that we were just providing new excuses for drinking and new ears for old yarns. Although we enjoyed the camaraderie in the beginning, by the time we left for the field we were happy to give our livers a break.

The Botswana Geological Survey was operated and staffed out of London by the British Overseas Surveys, the successor to the organization that had previously run the colonial geological surveys in the days of Empire. There wasn't, in fact, all that much difference between the new overseas surveys and the old colonial ones. Many of the technical departments in the Botswana government, such as Lands and Surveys, the Geological Survey, and Wildlife, were still managed by the British, as in many of their other former African colonies, because there was not yet enough indigenous technical expertise for Botswanans to take over those functions. The director of the Botswana Survey, Dr. John Hepworth, was an old Africa hand, having spent more than thirty years with the colonial and later overseas surveys of Uganda and Tanzania before taking on this directorship, which was to be his last overseas posting. The professional staff consisted of eight or ten younger geologists and geophysicists, recruited from Britain, and thirty or so Afrikaner water-well drillers, who spent most of the year in camps all over the country boring wells for villages, mines, and cattle posts.

Dave showed us around the survey. Their main duties, other than the water-well drilling, was in assessing the country's mineral resources. The geology of Botswana is very similar to South Africa's, and its mineral resources promised to be just as rich. Large copper-nickel deposits had already been found at Selibi-Phikwe and near Francistown, and the kimberlite fields recently discovered near Orapa promised to become one of the world's largest diamond mines. These deposits were the promise of the future for the young country, and the Geological Survey had the responsibility to advise the government about their proper management. The young geologists working there were very keen, and had plenty of money for modern equipment, like computers and geophysical prospecting instruments. Their main problem was that they were hampered by a lack of technical infrastructure. There was such an educational gulf between them and their Batawana staff that everything of the slightest technical nature, from soldering to computer programming, had to be done by them, overtaxing their time and technical skills. As a result, a substantial fraction of their instrumentation was out of commission.

Since we had plenty of time on our hands waiting for our camp to be organized, Teddy pitched right in and started repairing various bits of their gear. He dug into this work with zest, since the one thing he hated more than being idle was to see equipment unused because it was in poor working order. It wasn't long before instruments that had been put aside for months, with the expectation that they would have to be shipped to South Africa or Europe for repairs, began to function again. He soon had the whole supply department busy phoning Johannesburg and running up to Gaborone to get all manner of replacement parts. They weren't used to ordering small component parts for repairs, like chips and capacitors, but Teddy quickly found the electronic supply houses in Jo'burg that stocked what was needed. He became instantly popular with the staff, with everyone bringing in their long-abandoned pieces of gear for repair in Teddy's electronic version of a triage clinic.

In the meantime, I had plenty of time to dig into Hepworth's extensive personal library on African geology, which he had graciously made available to me. There I found a rare copy of Siegfried Passarge's *Die Kalahari*, published in Berlin in 1904. Passarge had been a geographer-naturalist in the mold of Alexander von Humboldt, and, during the last decades of the nineteenth century, had been the first European to extensively explore and map the Kalahari. His book contained a beautiful map that showed that in those days the Mababe Depression, northwest of the Okavango, had received water from both the Chobe and Okavango systems. It also contained his reconstruction of the earlier drainage system of the Okavango into the Makgadikgadi salt pans, before it was interrupted by the development of the Okavango Delta. I also found a history of Livingstone's early explorations of the area. On his first trip into the African interior in 1849 he had crossed the Kalahari and reached Lake Ngami, which he described as being seventy-five miles in circumference. It is less than half that size now, though its former shoreline could still be seen on my satellite images. During Livingstone's era, the southernmost channel of the Okavango Delta was its principal outlet, flowing into the lake, which then had as its outlet the Nhabe River, which flowed north. In the 1870s the flow from the delta into the lake began to diminish, and the Nhabe River was observed to flow in different directions at different times of the year. By the early part of this century the lake's former inlet had become completely blocked, and its former outlet became its principal inlet. Here was further evidence for the instability of the area, and for my thesis that tectonic activity was a contributing factor.

I had only a brief interview with Hepworth, since within the fortnight he was leaving for England and retirement and was busy getting his things in order and breaking in his replacement. He came into his library one day and pulled up a chair opposite me at the big table where I was poring over a stack of old reports from the Uganda survey. "Sorry I haven't had a chance to have a chat with you. I must apologize that I have only the foggiest idea of what you're up to. Mind filling me in?"

I sketched out for him the gist of my hypothesis regarding the Okavango and the idea of the microearthquake survey that I thought would provide the critical evidence.

"Hmm. Yes, I see. Young Reeves had much the same idea. Quite possible. I think there are still a few arms of the rift yet to be found. They are quite indistinct this far south, you see. Yes, quite possible. Very difficult to prove, though."

I sounded him out on another topic, one that was of great interest to me—whether or not the present day rift system followed old, previously active zones of weakness. Were the present rifts new fractures, or were they the refracturing of an old, healed, but still weakened fracture system? This point is debated endlessly in the scientific literature on the African rifts but, being an outsider, I had not been able to come to a firm conclusion regarding it. I was interested in the opinion of Hepworth, who had spent a lifetime studying the rifts but had actually published very little about them. He was one of that brand of geologist that sticks quietly to his own work and does not enter the arena of grand public debates that swirl around in the scientific literature. That kind of person often has very sound opinions that are not prejudiced by his having taken public stands on an issue.

"Depends on what you mean, I suppose. McConnell has been making a big stir of late claiming the rifts follow Archaean transcurrent shear zones. Pure moonshine! There are just as many places where the Archean structure is at high angles to the rifts as parallel. In any case, you usually can't even define a shear zone in Archaean terrain, they are so thoroughly sheared up everywhere. There is something to be said, though, about the rifts generally being controlled by Precambrian structures. They never cut the cratons, see. There is many a place where you can follow a rift that is just cutting along an old mobile belt on the edge of a craton and that never runs into the craton itself."

When I went over my field plan with him, he expressed some reservations. He put his finger at a spot on my map where I had planned a base camp in the Chobe National Park. "Why did you pick this particular place for your camp?" he asked.

"Well," I said, "it is central to these kopjes where we need to install our instruments, and it looked like a pleasant spot, with a bit of shade and water."

"That is exactly the problem with this spot. Do you see here, just at the end of this little stream that you want to camp by? That is the biggest water hole in the central Chobe. There are anywhere from one to five thousand animals at that hole every night during the dry season. In fact, the Ministry of Construction had a party at that site five years ago. They were supposed to build a bridge across that stream, but they never finished. The crew mutinied because of the animal activity. As a matter of fact, we haven't operated a party in the Chobe ourselves for the last five years. It's considered too dangerous."

"Oh," I said, in my tiniest voice.

"In any case, you'll need special permits to work inside the park. I'll ring up the Ministry this afternoon and get your party cleared."

"Thanks," I said. It was clear that my interview was over. In any case, at that moment I didn't feel like bringing up any other aspects of my plans.

Dave went to "Gabs" for the weekend to attend a birthday party for one of his mates, and Teddy and I accepted his most welcome invitation to move from the Cumberland to his place. He lived in a small bachelor bungalow, which was very sparsely furnished and had plenty of extra space for us. Teddy won the coin toss and got the bedroom, and I got the sofa in the living room. Dave had a young domestic, whom he called "girl." She was either very shy or afraid of us and spent the entire weekend hiding in her hut behind the house. It was several days before I realized that Girl was her name.

That evening we went to a *braii* at the Lobatse Club. The recently completed clubhouse was the most impressive structure to be seen in Lobatse. Its semicircular plan contained a spacious air-conditioned dining area and lounge, and alcoves for cards, darts, and billiards, all facing on a rear glass wall that opened onto a garden of shockingly green grass, possibly the only existing lawn in the country other than the patch at the Holiday Inn in Gabs. There were even plans for a swimming pool. It was clearly designed to be the centerpiece of the white society of Lobatse. The status of the Cumberland Hotel seemed destined to slip another peg or two.

The *braiivlace*, a southern African barbecue, was laid out in the garden. On the grill were chicken, steak, roebuck, and Portuguese sausage, accompanied by mealy-meals and salads of various types. Practically the entire white community was there that evening and since we, as newcomers, were novelties, we met a lot of people whose names I kept mixing up for the rest of our time in Lobatse.

We were also novelties as Americans, since everyone else was either British or white South African, working for the meat company or in management for the various technical agencies of the government, such as the Geological Survey. All in all, it was an enjoyable evening, although midway through I got cornered by a pinched-faced sourpuss of a woman, a Mrs. Gladys Davies-White, who seemed to take it upon herself to vet me on my personal habits and predilections. It started with her inquiring about what activities I might be interested in. She seemed to be interested in fitting me into the club social schedule, which was extensive. There was the round-robin tennis tournament, just starting up, or the knock-out bowls tournament in a week or two, or perhaps I might enjoy the bridge club? "Everything is going to be so wonderful now that we've got the new clubhouse opened, don't you think?" It seemed there was some activity or other scheduled for every afternoon and evening of the week. I suspected that Mrs. Gladys Davies-White might be the leader of a wives' clandestine Anti-Cumberland Bar League. Since I was hoping not to be in Lobatse any longer than necessary, I was not very responsive to what she probably considered her warm hospitality. Eventually she got off the activities topic and began telling me various bits of scandalous gossip, pointing out to me several of the women present who, according to her, had unsavory reputations and whom I should avoid. I pricked up my ears at that part, but just as she began relating a particularly juicy morsel I was rescued by Teddy, who called me to join him in a game of darts.

In the wee hours of Monday morning I was awakened in my sleeping bag on the floor of Dave's living room by the lights going on and then being tripped over by a young black woman wearing what I first took to be a Day-Glo pink bikini but which turned out to be her underwear. Teddy came grumping out to join me from the bedroom from which he had been evicted by Dave. It appeared that Dave had made a new acquaintance in the Holiday Inn and, forgetting he had houseguests, had brought her home. We were awakened again in the early morning by Girl, who seemed not to have taken kindly to the new domestic arrangements and was having a terrific row with Dave in the kitchen. We crept off tactfully to the Cumberland for coffee and breakfast, meeting Dave again mid-morning at the survey. He looked a little seedy after putting his new friend on the morning bus back to Gabs. "Sorry, you guys," he said sheepishly. "All I can say is, it seemed like a good idea at the time." It was classic.

We had to run up to Gabs ourselves the next day, to settle some unfinished business. We stopped in at the U.N.D.P. office, where our paperwork was finally ready, and at last made contact by telephone with Dr. Ernest in Maun, who let us know that our bungalow, which we still had no intention of using, was now ready for our arrival. More important, he told us that a Land Rover and driver would be provided us and that the sixteen new car batteries that we had requested had arrived. This was a particularly good piece of news, because they would provide the main power for our instruments, and lead-acid batteries are always hard to obtain in quantity, even in good-sized towns in developed countries. We probably would have had to go to Jo'burg to get them if they hadn't been delivered to Maun. The accountant put us through some more rigamarole about our travel allowances. She asked us if we were going to be posted in Maun or Gaborone during our visit. We told her that we would be in the bush, not in any town. "I guess that is `elsewhere,' then," she said, checking off a box on her form. "In that case you will each get ten rand fifty per day for subsistence." I explained to her that we already had our expenses covered under our grant, but to no avail. According to the documents we had picked up in Rome, we were to get per diem allowances. And that was that. I was a bit upset by this bureaucratically mandated double-dipping, but Teddy was quite pleased, considering it something like the "sea pay" he had been used to as a seagoing technician. "Something to take home to Judy," was the way he put it.

We next stopped by the Land and Surveys Department. I wanted to find out if there had been any repeated leveling done on the road to Maun, from which I might be able to glean some data on recent tectonic movements. The office was run by a bunch of Brits, slickly groomed and all dolled up in freshly pressed shorts, knee socks, and white cricket sweaters. They looked like fashion plates for some high-toned tropical outfitter's catalogue. When I made my admittedly overly hopeful request, one turned to the other and said, in tones used between the OK people when speaking about the non-OK people, "Did you hear that, Richard? These people actually want to know if we've ever surveyed a road twice. Sorry chaps. You've come to the wrong place. Country, actually. Haw-haw."

"What a snide sonofabitch," I said to Ted as we left. "I bet those pricks have never been on a survey party in their lives."

The office of Air Botswana again professed no knowledge of our shipment or when it might arrive, so I sent a cable to the shipping department at Lamont asking someone to trace it from New York. The next day we received their reply, informing us that British Caledonian Airlines had delivered our shipment to Lusaka the previous week. I called Zambian Airlines in Lusaka and they claimed not to have received it. Things were definitely not looking up.

Jake dropped by the Cumberland after lunch the next day and invited us to the golf club. This was possibly the world's first ecologically harmonious desert golf course, a nine-hole course carved out of a thorn and scrub thicket with rolled dirt fairways, macadam "greens," and plenty of unsculpted sand traps drifting across the fairways. The clubhouse was a concrete block shed under a big tree. Inside, a rattan bar was set up in one corner where we found, inevitably, a group of Jake's cronies. After a few beers I felt the call so I wandered down to the tin outhouse alongside one of the fairways. While I was inside conducting my business there was a sudden earsplitting roar, shaking the whole privy and filling it with dust, old cobwebs, and other debris I would rather not have been breathing. Holding the top of my jeans together, I bolted out the door, to see the tail end of a C-130 Hercules cargo plane receding down the fairway about fifty feet off the ground. Shaken, I ran back to the clubhouse, where everyone was standing outside, watching the giant plane wheeling about for a return run.

"What the hell was that?" I sputtered as I struggled to button up my fly.

"Oh, that vas just the Air Alaska boys," chuckled Jake. "They yoose the third fairway as a landing strip, so they alvays buzz it first to get everybody off. Gave you a bit of a fright, though, eh?"

"No shit. I thought somebody had blown up the outhouse!" We watched the big plane come in for a dusty landing. "What the hell is Air Alaska, anyway?"

"Like maybe they got lost? Heh-heh."

"They're a freight outfit. Yanks. They get most of the meat company's contracts for air shipments."

"Oh? Do they fly into Lusaka?"

"Sure. They fly everywhere they can get cargo, those boys.

Rhodesia, Mozambique, Angola, Zambia, wherever, hey? Not too particular about what kind of cargoes they carry either, from what I hear, nor does anybody ask them. All those countries are at war, one way or the other, but it's in everybody's interest not to know too much about what goes in and out of that Herk."

Teddy and I looked at each other, with a "let's go" connection. We ran to the car and took off down the fairway in the direction the plane had disappeared in. We reached the end of the "landing strip" just as they were shutting her down. The rear door was already lowered, revealing the cavernous cargo space half full of crates strapped to freight pallets. Half a dozen cars and pick-ups came out to meet the plane from the other direction. Two pilots climbed down from the plane, which was emblazoned on the side "Air Alaska International." They were decked out in fancy airline pilot uniforms, with enough gold braid to decorate the sleeves of the commodore of the yacht club. We waited until they had greeted their families and members of their ground crew, then approached them and explained our problem with our missing shipment in Lusaka.

"I believe we've seen it. There has been a shipment sitting on a pallet next to the freight office in Lusaka for the past week. We wondered what it was. But if it's consigned to Zambia Airlines you may be waiting quite a while for it. They only run two flights a week to Gabs, DC-3's, and they're usually full up. They would have to bump passengers to put your shipment on board, and they won't want to do that. They've already got your money, see."

It was the good-news/bad-news routine. "Are you guys planning to fly to Lusaka anytime soon?"

"As a matter of fact, we have a run on Thursday with a stop there."

"Can you take a passenger?"

"It wouldn't be very comfortable. All we've got is a hammock strung over the rear freight door."

"Hmm. What d'ya say, Ted, feel like taking a little ride on the Reading?"

"Heh-heh-heh. What the hell. Anything for a good cause," Teddy replied, with his nervous laugh that signified: "Holy shit, things are starting to get a little out of hand here."

We finally agreed that the best move would be to give them a copy of the waybill with a letter authorizing them to take over our cargo for trans-shipment to Botswana.

That evening Ted went off with Dave to carouse somewhere in the native quarters, and I went to a party at someone's house. I met one of the Air Alaska pilots there, who filled me in about their airline. It had been formed by a group of pilots from "Air America," the C.I.A.'s house airline in Vietnam. They had bought several C-130s from the Company when the airline was disbanded at the end of the war and set up shop in Alaska, transporting oil exploration gear during the Prudhoe Bay development. When that started to wind down, they spread out anywhere in the world where they could land freight contracts, like the one they had here with the Botswana Meat Company. Most of the pilots and ground crew had been together since Vietnam. This pilot, at least, didn't seem to have the slightest worry that every day they were flying in and out of at least two war zones. "War means more freight business, and with the railroads shut down the way they are, that means air-freight. And we've got the planes."

Later, while sitting on the floor eating my buffet supper, my enjoyment of the meal was ruined by a conversation going on behind me that was pretty hard to ignore. It was my old pal Mrs. Gladys Davies-White, sitting in a rocking chair, knitting and gossiping with some of her cronies. They were cattily trading news and rumors of various lurid scandals. To listen to them you would think that every white man in Lobatse was sleeping with someone else's wife, had a black mistress, or both. Fortunately, I was shortly joined on the floor by a young lady who was friendly and outgoing, and we started up our own conversation. Her name was Doris, and she had come out from England a year or two before on a three-year contract as a secretary for the Geological Survey, where I had met her briefly before. We were just trading pleasantries at first, but she had had a bit too much to drink and soon began a maudlin soliloquy about how unpleasant it was to be a single woman in a place like this. It seemed she had originally come down to Botswana to join her boyfriend, an accountant for the wildlife service, but by the time she had arrived he had already become engaged to someone else, a South African woman he had met on a trip to Jo'burg. Since she couldn't break her contract she had to stay on, stuck in a very uncomfortable situation. As the only single white woman in Lobatse, where the men were either married or totally footloose like Dave, she was constantly the subject of rumors, whether she had affairs or not.

As if prophetically, a sharp voice from behind me said, "Well, Doris, I see you've met the new man. Not wasting much time, are we?"

"Hello, Gladys. Why don't you mind your own business?" was the equally icy reply.

"By the way, dear, where's your darling friend from the townships, Chief what's-his-name? It's terrible, I can never remember the native names. It's so much easier when they have English names, like `boy.' "

Doris gave her a deadly look, and tried to resume our conversation, but, ignoring me, Gladys and her henchwomen wouldn't let go and proceeded to slice and dice Doris with their sharp tongues. Doris, had she enough confidence and wit, might have been able to defend herself, but in her present condition she was no match for their vitriol, and was quickly reduced to a blustering but teary wreck. I felt terribly sorry for her, as well as very uncomfortable myself, and was ready to suggest giving her a lift home, whatever the consequences of that may have been. Just then, the door opened and in stepped a very handsome and elegantly dressed black man. Doris seemed suddenly transformed and greeted him warmly, her eyes suddenly taking on a deep glow. The man, quickly sizing up the situation, courteously addressed the group of harridans with a graceful little bow and polite greeting, and swept Doris out the door, followed by a chorus of "hmmphs."

Such was the nature of this tightly knit and stressed community of expatriates that Teddy and I were drawn into its bosom with the welcome of true believers into a cult, or fellow addicts into a drug den. We were white, and that was enough to be accepted as fellows in their struggle to maintain what they considered a semblance of civilization in this outpost. It was a down-market colonial enclave, a vestige of a long-departed and no longer profitable empire. Here there was no Raj, just a bunch of Tsetswana tribesmen to lord it over, but who were in any case living in their own sovereign nation. These "colonials" were no more than middle-class provincial Englishmen and women on civil service contracts, putting on airs and sporting a few servants. If my descriptions seem like caricatures, it's because they acted out parts and recited lines that might have been written by Somerset Maugham on a bad day. There was a cartoon from Punch tacked up in Dave's office. It showed a couple of British explorer types, in shorts and pith helmets, standing in front of the proverbial graveyard of the elephants. The caption was, "Gad, they'll never believe this back in Torquay-Winton." I got a secret howl out of this, not because it was so funny, but at the thought that in pinning it up, Dave hadn't fully appreciated its irony.

Trapped as we were in this world apart, Teddy and I did not have much contact with the life of the greater part of the population of Lobatse, but the few glimpses we got had a certain charm. On our first Saturday, we were disturbed from our post-lunch lassitude by a hubbub out behind the Cumberland. Ted investigated, and returning, shouted, "Hey, they're gonna show a movie. It's Saturday matinee time. Come on, let's go."

And indeed it was. Around back of the hotel there was a big quonset hut sort of a building, with a big crowd milling outside, about half schoolchildren, half adults, all black. We filed into the building, which was set up like an auditorium inside, and everyone took a seat on folding chairs, with the little kids sitting in the front on the floor. Someone started up a 16mm projector in the back and everyone quieted down. The first feature was some sort of "News of Britain" short. It began with Queen Elizabeth greeting some dignitaries at Buckingham Palace, which produced no particular reaction in the audience. This was not at all the case with the next piece, which featured the races at Ascot. The opening scene was a head-on view of the horses coming out of their posts directly at the audience, which screamed and pulled back in panic. Before the audience was in full flight, the scene switched to a side-on pan of the horses racing past. The audience, relieved and still thrilled, jumped and shrieked and shook their arms at the screen. It was bedlam for the next few seconds until the scene switched to the toffs in the grandstands, and the audience suddenly quieted, as if to say, "What happened to the horse race?" This must have been what it was like when the Lumière brothers screened "Arrival of a Train" in Paris in 1896, to audiences that were terrified they were going to fall under the wheels. Coming from a country where children are jaded at age three by television, witnessing such cinematic innocence was a completely new experience for me.

On the next day, Sunday, we were left to find our own amusements. We decided to take a drive out to the west of town to explore an alternative route into the Kalahari that was on some maps but evidently was no longer used much, since we had been able to find no one who could tell us about it. The first part, out as far as the tribal village of Kanye, was a well-maintained graded road. On the way to Kanye we met an astonishing sight, a long train of wagons moving toward us along the road. There were covered wagons in the train, big Boer trekker wagons, with their canvas covers cinched down in the middle like Conestogas and pulled by long teams of oxen. There were donkey carts, and big solid-wheeled oxen-drawn drays. They were filled with the Kanye villagers, all dressed up in elegant finery as anachronous as their transport: long white dresses and bonnets for the women and black frock coats and tall hats for the men. It looked as if they had outfitted themselves at a Voortrekker rummage sale in the previous century. They were on their way to a big wedding at another village. There they would stay for a week or two, celebrating with their distant cousins.

When we finally got to Kanye, it was deserted. It was a beautiful village in a very graceful setting: thatched rondavels in spacious thatch-walled compounds scattered, with generous space between, in a region of low rolling hills studded with acacias. Here there was none of the crowded squalor of the native quarters in the towns. We wandered through, stopping in the central square, with its central water pump opposite the schoolhouse. What was striking was how clean and tidy the village was. There was not a bit of rubbish to be seen anywhere. The hard-packed dirt of the compounds and public paths and squares had been carefully swept. The good women of Kanye had left an impeccable village that would pass muster, during their absence, with the most critical of passers-by.

5

There was finally some good news for us when we met the Air Alaska flight on Thursday afternoon. The chief pilot, Duncan, walked over as soon as he spotted us waiting among the usual welcoming party. "Your shipment will be on tomorrow's regular Zambia Airline flight to Gabs."

"Great! What happened?"

"Oh, no problem. When we showed them your paperwork, those boys jumped to attention. They don't want to lose their piece of the cartage, see. No fear, it'll be on tomorrow's flight. We watched them put it aboard ourselves."

The next day, sure enough, it arrived in Gaborone. We brought it back to Lobatse and unpacked. Our little expedition was now firmly scheduled to depart first thing Monday morning, so we had a couple of days to get everything checked out, put in working order, and repacked for the field.

We spent the last couple of days shopping for all sorts of odds and ends that could be obtained locally. Small batteries, hand tools, and various types of spare parts and personal camping gear were on our list. Dave had also told us to stock up on plenty of chloroquine, the malaria prophylactic, and get a snake-bite kit each. The snake-bite kits turned out to be small boxes the size of first-aid kits and cost fifty rand a pop, which surprised me, since I was somehow expecting the little units with the rubber suckers and string we get in the States. These we threw in the lockers under the front seats of our Land Rover.

On the Saturday before departure, we drove down to Mafeking to get food stocks. The only food that could be reliably purchased where we were going was meat and mealy-meal, so before leaving we had to buy anything else we might need for the next three months. Mafeking was a disappointment, considering its fame in the history books. It looked much like any run-down cowtown that one might find in Montana or Wyoming, with high railed sidewalks lining dusty streets. In the park in the central square was a statue of Baden-Powell, the "hero of the seige of Mafeking," of British historical legend. When we remarked on it, Dave scoffed, "Yeah, the boy scout. The Boers cut Mafeking off but never actually tried to attack the place. All Baden-Powell did was organize social activities and compose heroic dispatches for London."

"Surely you didn't learn that in a schoolhouse in Dorset?" I remarked.

"God's truth, but I've been doing some reading since I've been down here."

We each got our own cart in the supermarket and loaded it up, both with staples and with any other special treats that we might imagine desiring over the next few months. I actually saw Dave slip the canned components for a complete British version of a Christmas dinner into his cart, up to and including the plum pudding and rum sauce. When we got to the check-out line we each peered into the other's cart to see what he had picked up. "What's that?" said Dave, pointing to half a dozen cans of coffee in my cart. "Coffee." "But I already got the coffee," he said, holding up a jar of Nescafé. "Dave," I said, "in the States we have this powdered stuff called instant tea. How would you like to be drinking that?" The Nescafé went back on the shelf. Over the next few months, as we delved ever more deeply into our food stocks, we would each gradually discover many strange and peculiar items that the others had selected during that mad hour of shopping in Mafeking.

Bright and early on the following Monday, we found our caravan being assembled on the main street in front of the Geological Survey compound. By eight o'clock in the morning, this had already attracted a large crowd. The wives and families of the crew had come to send off their men, and various relatives, friends, kids, and casual passers-by attracted by the unusual activity formed a crowd that stirred up a cloud of dust and blocked the whole street. There were five vehicles in all, lined up along the wall of the compound. There were two long-wheelbase Land Rovers, a couple of F-250 Ford pickups and, at the center of attention, a five-ton Bedford stake truck that was surrounded by a huge jumble of gear that was being sorted through, handed up, and stowed away by a gang of laborers. The bed of the Bedford contained our vital fluids, sixteen 55-gallon drums, half of them filled with water and the other with petrol. We watched with growing amazement as the contents of our African bush camp were loaded up on top of the drums. Tents, cots, tables, chairs, mattresses, lamps, cooking gear, a bathtub and a shower, and a host of unrecognizable items were hoisted up and carefully tucked away until the truckbed was filled right up to the top of the cab. The camp laborers rode above this load, sheltered under a canvas cover in little nests they made in the stowed bedding and made comfy by pots of food and parcels of personal gear hoisted up by ropes from their families on the ground. Meanwhile the scientific equipment, the more delicate items, and our personal gear were being distributed among the other vehicles and carefully stowed away. All of this was accompanied by a hubbub of excited jabbering in English and Setswana, as the crew bantered hilariously among themselves while the crowd kibitzed with shouted encouragement and advice to the men loading the trucks. By now everyone was infected with the "leaving for the field" spirit.

Teddy and I were astonished at this spectacle. Up to this moment we had not actually known what our often-discussed "camp" was going to consist of. What we now saw was beyond our wildest dreams. The amount of gear loaded up that morning looked like enough to make up a full-fledged safari camp from back in the Great White Hunter days of the '30s that I recalled dimly from anciently popular novels of generations past, found in seldom-visited recesses of the public library. When the loading was complete, there was a little ceremony in which the crew was lined up before the vehicles and introduced to us by Dave. They were eleven: Blackie, captain of the five-ton and boss of the crew; four drivers; five driver's assistants (laborers); and Deacon, who was our campkeeper, a euphemism for personal cook-valet.

"Not too shabby," said Teddy. "Heh-heh-heh."

It was midmorning by the time we had loaded up and gone through countless permutations of farewells and last-minute delays, like the retrieval of a special stew cooked for a departing husband the night before and forgotten on the coals. As a result we didn't reach Gaborone until noon. We took a break there, because Blackie had to pick up a stock of government petrol chits and a load of survey gear to be delivered to Francistown. Dave, Teddy, and I went off to the Central Hotel for lunch, where by chance we barged in on a farewell luncheon for John Hepworth, and joined in the festivities. Somewhere in the course of this celebration, Hepworth pulled me aside to offer a piece of advice. "Good luck to you on your project. We've given you a very good crew, so you should be well taken care of. Just remember, though, they are all from the south of the country and have never before seen wild game, let alone elephant. So they are in your hands when it comes to dealing with problems with animals. You can't be too cautious, particularly with elephant. If you have any problems, don't be afraid to pull out. Nobody will blame you in the slightest." This was good advice, of course, but like most good advice, it is not so easy to apply in practice. By chance, I was to meet Hepworth once again. A few years later we ran into each other near midnight on the steps descending to the Fellows' Bedrooms of the Geological Society in Piccadilly, where we both happened to be staying at the time. I gave him a drink in my room, and a brief recounting of the scientific results of our project. He seemed genuinely interested, and never once asked whether we had encountered any difficulties in the field.

Heading north out of Gabs, we entered a region subtly different from the southern part of the country around Lobatse. We were on a broad arid plain, very gently sloping off to the east. Scattered dilutely over the landscape were areas of dry grasses and mesquite-like brush separated by large barren spaces of unstable soil and sand. There were few entrenched watercourses. You could see that when it rained here in the season it came in torrents, in floods so fast that the water just sheeted across the desert, leaving wide regions of braided streambed that did not flow long enough to erode down to any depth. In contrast to the southern terrain, here kopjes could be seen only dimly on the horizon to the north and east. Far off to the west a low row of hills separated this plain from the Kalahari proper.

The road, unimpeded, stretched out straight to the horizon. This was the main highway of Botswana, paralleling the railway that lay ten miles off to the east. Graded out of the desert floor, the road was just wide enough to allow two trucks to pass at speed. Its surface was firm, but sometimes it ran on beds of calcrete that would unexpectedly change from hardpan to a soft powder, throwing the pickup into a violent lurch with a jerk of the steering wheel that threatened to break the driver's wrists if he didn't lay off the gas fast enough. Our caravan was spread out over a distance of two miles to let the dust settle between vehicles. Pity the few drivers who had to pass us going in the other direction.

The pickup I was in was bringing up the rear, so I noticed by the dust clouds that our party was coming to a halt way up ahead. When we pulled up behind the line of parked trucks, Dave and Teddy and some of the crew were standing on the other side of the road around a small stone monument. It had a brass plate mounted on it, on which was inscribed:

Tropic of Capricorn
23° 30' South
The Southernmost Declination of the Sun

There was a small cavity in the side of the monument, on the top of which was a narrow slit that admitted sunlight. Immediately beneath the slit a line was scribed. It was September, so the bright line of sunlight coming through the slit fell to the south of the line. It would fall directly on it on the winter solstice. We took pictures to record the occasion, and Teddy and I did a little Alphonse and Gaston number, welcoming each other to the Tropics.

Late in the day I saw my first baobab tree, and made the driver stop. It was an immense solitary, silhouetted in the dust-sunset of the desert afternoon. Its enormously thick bole grew straight up for ten or twelve feet, then abruptly terminated in a crown of thick limbs extending outward and upward and fractally dividing into finer parts like some kind of giant subaerial coral. Its form is unlike that of any other tree. Disdainfully, Livingstone had described it as "a carrot growing upside-down." In the top of its trunk at the center of its crown of limbs there is a bowl-shaped depression that fills with water during the rains and serves as a cistern both for the tree and for animals like leopards that climb it to lap the water. I later learned another fact about baobabs that would be particularly useful to us: They grow only where they can anchor their roots on rock, so wherever there are baobabs, bedrock lies near the surface.

When we returned to the truck, Samuel, the driver, told his assistant Robert to refill it with petrol. He knocked the bung out of one of the drums kept behind the cab and inserted a garden hose into it. He put the other end of the hose in his mouth and gave a powerful suck, pulling a siphon on the hose which he deftly removed from his mouth and inserted into the truck's filler hole, all without swallowing any or spilling a drop. This was the crew's standard method. I had to do it once and spent the rest of the day trying to get the taste of gasoline out of my mouth.

Five or six long hours out of Gabs, we rolled into Mahelapye, a village of huts clustered along the edge of a dry river gorge. There was a good pub there, since it was the only stopover point between Gabs and Francistown. It consisted of a compound of thatched-roofed rondavels surrounding a central building with a small restaurant and bar, perched on a high bluff overlooking the town. We checked in while the crew set up camp in an open space outside the compound.

After dinner we stopped by the bar for a game of darts and a beer. We took on one of the local dartboard habitués. He played with a cheap set of darts with ragged plastic flights that he kept stuck in his greasy afro when he wasn't throwing. But he knew his way around the board and neither of us could come close to beating him. The room was completely full with about a dozen locals and a similar number of travelers. We ended up at a table with a bunch of salesmen of agricultural goods who were having an animated conversation about pumps and well-lining equipment. A couple of them started reminiscing about the big rains earlier in the year, when they had been marooned in this same pub. Mick, a ruddy-faced giant of a man with a gut that must have been a wonder to see him prize behind the wheel of a Land Rover, proceeded to fill us in on the story.

"Yeh, yeh. That was the biggest rain we've seen for 10, 20 years. It was the only time I've ever seen the river here flood. Right up to its banks it was, too, and right over the top of the bridge. There must have been a foot and a half of water over the bridge. That's why we were here, y'see. This was the only place for miles around that wasn't under water. And a'course, nobody would dare cross the bridge in those conditions. It's bad enough when it's dry. Everybody heading for Francistown was stuck here on account of it. We were stuck here for two days ourselves, and by then it was chock-a-block full, boys kippin' in the kitchen and in the bar. But, hell, we just made a party out of it. Bob, the owner, kept the bar open late for the occasion. He was makin' a pretty penny out of it, for sure, and no fear the constable would show up and give him a citation, either."

Mick's mate Dilly chimed in, "Tell 'em about the daft Afrikaner, Mick."

Mick gave the table such a whack I had to grab for my beer before it went off the side. "The bloody Boer, how could I forget it! Now there was a case of a bloke that had the Afrikaner angel sittin' on his shoulder, if ever there was one.

"We were all sittin' in here, see, havin' a drink and feeling good about not being out in the pourin' rain. And it was right pissin' down out there too. Then all of a sudden the door there bursts open and in comes this bloke, lookin' right done in. He was soaked, his hair was all matted down over his face, and he had a wild look in his eyes. We all turn around and stare, thinkin', what kind of a dumb Afrikaner would be out on a night like that?

"So he sits down and I buy him a drink. 'Tough night to be out,' I says, friendly like. 'Bloody awful, bloody awful,' he says to me. 'Water everywhere. I couldn't even see the road for the last 40 miles.' So somebody asks him, 'What's the flooding like in Gabs, then? You must have been the last to leave there.' 'Gabs,' he says, 'I haven't been to Gabs. That's where I'm going.' We look at him like he was daft. 'Huh?' 'Well if you're goin' to Gabs,' I says, 'where in bloody hell did you come from?' 'Frauncestown,' he says. You could have knocked me over with a feather at that. 'If you came from Francistown,' I says, 'then how in the world did you cross the bloody bridge?' He looks at me, bleary eyed, and says, 'What bridge?'" Mick and Dilly exploded in bellows of laughter.

I didn't properly appreciate the story until the next morning, when I saw the bridge. It was actually the top of a narrow concrete arch dam thirty or forty feet high. The roadbed was just wide enough for a truck, had no guardrails, and was curved, following the upstream convexity of the dam. It would just take the Bedford with a few inches to spare on either side. Blackie had his assistant get out and walk in front to guide him as he inched his way across. The idea that someone could cross that in the night when it was covered with a couple of feet of water and not even know it was there beggared the imagination.

We got into Francistown about noon. It was another one-sided railroad town like Lobatse, being the last stop on the Rhodesian Railway until Plumtree in Rhodesia. Fronting the tracks along the main street was a bazaar of Indian shops, but here there also were curio shops, selling native carved wood and ivory handicrafts. Francistown was the largest town in northern Botswana, and the gateway to Maun and the national parks, so we were entering tourist territory. We took a leisurely lunch in the pub while the crew went off to the Geological Survey office to deliver the survey gear we had brought up from Gaborone and to exchange one of our pickups, which needed repair.

Leaving the main route on the outskirts to town, we pointed west for the long leg into the Kalahari and Maun. For the first hour or two to Nata we traveled through the same scrubland country that we had been seeing for the past two days. Even though prominently displayed on the map, Nata is just a locality. The only things there are a big baobab and a highway maintenance shed. For many miles out of Nata we traveled on a narrow rutted lane beside another lane that had been graded and covered with a fine white powder. When it rained, this would turn into calcrete, a sort of low-grade cement, provided, of course, it hadn't been blown away first by the wind or washed away by too heavy a rain. It was a clever and economical method of road maintenance, although to my tastes it seemed too precariously dependent on divine providence.

We left the eastern hills behind and entered a totally flat world. Here there were not even the thorn trees to interrupt the bleakness of the landscape. Grasses, brown and dry and no more that a few inches high, grew thinly over this plain, the only feature of which was the road, a scraped line that stretched ahead of us as far as the eye could see. Proceeding for miles down an imperceptible declivity, we found the sand being gradually overlain by salt. We were entering the Makgadikgadi salt pans, the lowest part of the Kalahari. In prehistoric times, tens of thousands of years ago, this had been the natural terminus of the Okavango River, where it had formed a huge shallow salt lake. This was before the recent tectonic activity had dammed it upstream to form the freshwater Okavango Delta.

We crossed mile after mile of salt beds, glittering in the sun. In places they were topped by boomerang-shaped barchan sand dunes, always on the move, that in good companionship slumped and slipped and drifted into one another. Here and there, in the ultimate depths of the playa, flat sheets of water shone from which, disturbed by our passage, flights of heron briefly rose.

As we approached the western margins of the salt, tufts of bright green marsh grass began appearing, growing on pilings of their own manufacture. We soon left the salt entirely, and passing though a dense fringe of marsh grass, entered a grassland. Off in the distance we saw herds of gemsbok and springbok, the springbok as they ran pronking high in the air in their characteristic stiff-legged leaps. This sight evoked in me my first feeling of wild Africa.

After passing through these grasslands for hours without seeing a tree, we saw palm trees in the setting sun, growing on the banks of the Boteti, the outflow river of the Okavango. Just at sunset, we climbed a low hill and abruptly arrived on the steep bank of the Thamalakane. Maun was spread out in front of us on the other bank, a maze of mud huts in mud-walled compounds, with a backdrop of forest in silhouette. Below us, in the river, were fishermen in dugout canoes. We crossed a rickety plank bridge and entered the town in the rapidly failing light. Our caravan drove through the dark town on a tortuous path through a labyrinth of roads that wound through a maze of compounds. Suddenly we came out in a square where a blaze of lights were centered on a complex of low whitewashed buildings. This was Riley's Pub, the last outpost on the edge of the Okavango. Our crew left Teddy, Dave, and me off there, with our bags and a Land Rover, and went off in the darkness to establish camp on the outskirts of the town.

6

The dining room at Riley's was a far cry from the Cumberland's. It also was staffed by a waiter wearing a red jacket and tassled fez over baggy trousers and taped-up sneakers, but otherwise it was unpretentious, with a dozen oilcloth-topped tables in a sunny room. We sat at a table next to a window where a group of weaverbirds were making their nests. They were slender, teardrop-shaped constructions, with tiny entrance holes near their bases, like slim delicate jai alai baskets. The bright yellow and green birds busily flitted back and forth, adding tiny fibrous bundles to their fine knitting. Whenever we stopped at Riley's in the ensuing months we always chose the same table to watch these birds, which never tired of adding finishing touches to their creations.

Botswana Air Safari Pty. was located in a modern block of tourist shops on the main street not far from Riley's. These posh shops traded in such things as zebra-skin rugs, leopard coats, and various other products made from animal parts, as well as finer examples of native handcrafted baskets and woodcarvings. They stood in marked contrast from the rest of Maun's mercantile district, a clutter of tumbledown shops dotted here and there haphazardly along the main street. The tourist trade in Maun was mainly limited to the well heeled—those in for the big-game hunting or on air safari tours—and these shops relied on them for their occasional sales. The air safari office, at the end of this row of shops, was austere, overly air-conditioned, and empty, except for a well-dressed, rather stately woman sitting behind a gray metal desk.

"Excuse me," I said, "we would like to arrange for a flight around the area."

With a measured gaze, she gave Teddy and me a quick but careful looking over, her lip curling slightly at the corner. "Not our usual customers," I could almost hear her thinking. "I'm sorry sir, but we are not booking tourist flights at the moment. There is a petrol shortage, and the government has grounded us for all but essential flights. Perhaps you might try back next week."

"But this isn't a tourist flight," I said, disconcerted. "We're on a government project. I've written to you about it several times in the last six months. The name is Scholz, from Columbia University in New York. It really is important that we get a flight today or tomorrow."

"I see. Just a moment." She knocked at the door to an inner office. A tall man in pilot's uniform emerged. "Hello, I'm Don Nichols, what can I do for you?"

I introduced Ted and myself. "We're on a Geological Survey project, starting up a geophysical survey of the area, and we need a flight to reconnoiter before we go into the field. I wrote you about it a few months ago. We need to fly the whole region from Lake Ngami to the Chobe to look for rock outcrops. "

"Oh, yes, I remember now. Right. Government project, is it, then? Well, if you'll just give me your government transport voucher, we'll fix you right up."

"I prefer to pay by travellers checks, if you don't mind," I said, pulling out a packet full of $100 checks. Who knows what it would take to get a government transport voucher, whatever they were. I certainly didn't want to try. I'd rather spend money than hassle with red tape.

"Right-io. U.S. dollars, hey?" he said, with a trace of a wink. "That'll be right, then. What you want takes about a two hours' flying time, so that will be six hundred dollars, altogether. Will one o'clock be all right? OK. Fine, meet me at the airport, then." That's what I like about pilots. They are always pragmatic. They like to fly and they have planes to maintain, and that's what counts with them. I didn't care whether he believed me or not, nor did I care how he got his petrol, for that matter.

We went looking for the U.N.D.P. office. It was located in a section of government bungalows planted in thriving bougainvillea. We found Dr. Ernest washing his car in the driveway of his bungalow. He was an Anglo-Indian, very brisk and officious. He took us around straightaway to the office, which was a prefab affair of similar design as the bungalows, but larger and with a different layout. Out front was a sort of bullpen of native workers lounging around in a collection of broken-down chairs and sofas, confabulating while waiting for work assignments. The interior contained one large room, crowded with desks and work tables, and several smaller offices in the back. In spite of the jumble of furniture in the workroom, there wasn't much work going on. He introduced us to a Swedish hydrologist who was fussing over core samples on one table. Across the room several men were perched at drafting tables chatting to several others collected around a water cooler. We poked our heads into one of the back offices, where Dr. Ernest introduced us to Dave Potten, the office manager, who was beavering away at a stack of paperwork.

"Welcome to Maun," he said, extending his hand in greeting. "Glad you finally made it. Grab a seat," he said, pushing a couple of folding chairs into a small clear space next to his desk. Dr. Ernest excused himself and left. "We've been expecting you for weeks. I hear your crew got in last night and are setting up a camp north of town. You could have had a bungalow, you know, but I hear you will be heading for the bush soon. Might as well get used to camp life, eh?" I immediately took to him. He was like a breath of fresh air compared to everyone else we had met so far in the U.N. organization.

"Not to worry, we have everything set up for you. Your batteries and charger are in the shed, and Thomas, your driver, should be here any time now with your Land Rover. Ah, here he is now. Come on, I'll introduce you."

A Land Rover had just pulled up out front. A short wiry man emerged, wearing a khaki shirt and a pair of enormous shorts from which his spindly limbs extended. On his head was a peaked woolen cap of some unrecognizably soiled plaid and, as if to further emphasize the scrawniness of his legs, he wore a pair of huge boots with no socks. A sly grin decorated his brown face, crinkly and of indeterminate age. He doffed his hat and said, with a vaguely military air, "Thomas Ingleton, sir. At your service."

"Well, I'm very glad to meet you, Thomas. I gather that you know this region well."

"I was born here, sir, and lived here all my life, except for when I went to the war," he replied, in what I now recognized as an oddly quaint English accent. I was to discover that this accent was restricted to certain stock phrases; otherwise he spoke the local pidgin.

"Oh, and where were you in the war?"

"All over, sir. I ran away when I was a boy and joined Monty in the desert. Ended up driving one of your jeeps all the way to Berlin."

That took me aback. "You don't say?" was all I could manage in reply. "Well, I'm very glad to have an experienced man with us, Thomas. I'm afraid no one else in our crew knows this area at all."

After arranging for Thomas to pick us up at our camp at eight the next morning, and collecting our batteries and other gear, we went off to meet Dave for a quick lunch at Riley's before our flight.

The Maun airport was nothing more than a single dirt strip, but it had a little terminal building on one side, with a waiting room with plate glass windows and a snack bar. This was provided for the use of the air safari companies, whose customers demand a modicum of comfort while they are out roughing it. Don was already there, in his little twin-engined Beechcraft, which he had already fueled up and cranked over. Not the most confident of fliers, I was always glad to see two engines on a bush plane.

We piled into the tidy little four-seater, took off and circled over Maun, heading south. From the air, Maun took on a semblance of a pattern not recognizable from the ground. It consisted of a thousand or so circular huts, clustered densely near the center and more loosely in the outlying districts that stretched for several miles along the banks of the Thamalakane. Woven through this jumble was an inosculating braidwork of roads taking almost every imaginable pathway. A few of these were graded and prominent, but most had been formed simply by repeated casual usage. The center of town was marked by a large dirt square, twice the size of a soccer field, fronted by Riley's and flanked by the post office, police station, and the block of tourist shops. The few other rectangular buildings were the shops scattered along the main street.

As we gained altitude, our view of the whole of this flat terrain expanded dramatically outward in all directions. A reddish-brown haze of blown dust mixed with the smoke of bush fires diffused from the land, blurring out the horizon so that the distant land and sky merged in an ocherous fuzz, topped by a bright blue cloudless sky, like a slightly out-of-whack spectral display. On our left the mottled reds and yellows of the desert were dotted here and there with the pale greens and olive of the sparse vegetation, which in the distance blended pointillistically into a uniform taupe. The palette was greener and bolder on the swamp side. Bright glittering silvery-blue channels could be seen here and there among the vibrant greens of thick growths of papyrus, lily pads, and other aquatic life that hemmed in the watercourses. Between these were sandy islands capped in thin forests of acacia and mopani. There was little soil developed in the swamps and the sands drained so well that semiarid conditions existed everywhere except in the immediate proximity to open water.

Flying south along the course of the Thamalakane, I tried to pick out the fault scarp that I had seen in the satellite images, but at fifteen hundred feet we were too close to recognize it as anything other than the straight left bank of the river. Dave tapped me on the shoulder and gestured from the rear seat. The pilot banked left in response. There we saw the disfluence, the place where the river split into two, with half the flow following the channel of the Boteti out to the salt pans, the other diverging at right angles and flowing down the Nhabe River to Lake Ngami. The Boteti had cut a narrow gorge through the scarp, showing its dissected profile. We descended and circled for a better view. Over the roar in the cockpit I gestured and mouthed my interpretations to the others, who nodded and gestured back.

In a few minutes we were over Lake Ngami. The surface was mottled in brownish and garish blue-green algal mats, slurried and swirled together like thick iridescent oil slicks. As we descended a spattering of white and pink flecks became vast flocks of heron and flamingos, feeding in the shallows where the water was pink with brine shrimp. Turning east at the southern tip of the lake, we flew on a fixed bearing I had passed over to the pilot on a slip of paper. We were trying to spot several groups of kopjes east of the lake, and since our maps were so poor, I was trying to get as many compass headings as I could. Soon the first group of these kopjes, the Kgwebe Hills, appeared out of the haze. From a distance they looked like a clump of shiny turtlebacks half buried in the sand. As we approached they became rougher and more complex in their connections and structure. Canyons with trees and underbrush, dominated by baobabs, became discernable. Suddenly we were over the top and we could see that the hills surrounded a beautiful little sheltered valley, lush with vegetation. Teddy gestured to the right. We could just make out what might be some huts among the trees, and some cultivated fields. Later we would wish that we had taken a better look at those, but just then we were more interested in rock than villages so we continued east to the next kopjes. These turned out to be much farther to the east than indicated on our maps, and were probably too far away to be of use to us in our survey, since they would be more than a day's round-trip by Land Rover from Lake Ngami, where we would be setting up our base camp in that area.

Turning to the south, we picked up the two lines of Wildlife Control Fences. These infamous fences were supposed to prevent the spread of hoof-and-mouth disease from the wild to the domestic herds, but they also had the tragic effect, described so poignantly by the Owenses in *The Cry of the Kalahari*, of preventing the wild herds from migrating to the swamps in periods of drought, causing thousands to perish against the wires from thirst and starvation. We took a bearing on them before crossing the Okavango diagonally on the next leg of the flight.

For more than half an hour we flew over the swamps, an impossibly complex filigree of streams weaving through sand, with pale yellow islets surrounded by lush channels, like amber beads set in malachite. Here and there we saw hippos wallowing in pools, and herds of antelope in meadows along grassy banks. Don gestured down to the left, pointing out a herd of elephants. He dove the plane at them and as we passed over at treetop level a big bull reared back on its hind legs, waving its forelegs and trunk at us defiantly. At the north end of the swamps we turned east again, and following a line of pans, flew out over the open veld of the Chobe National Park. The pans looked like splat marks from the air, like meteorite impacts on the moon. They are circular white areas surrounded by rings of trees from which a starburst of white lines radiate in all directions. Pans are shallow depressions in which water collects, and the radial lines are the animal trails leading to them. They often are strung out in lines, like the one we were following, because they were originally localized in the troughs of old linear sand dunes.

The Chobe is a huge expanse of dry veld, covered in places by dry and sparse forest. It is entirely a wildlife refuge, with no human habitation except for a couple of small game scout camps. It stretched out now before us as far as we could see to the north and east. We began to see concentrations of animals, mixed herds of antelope and more and more elephants. This was partly because they were being herded toward us. Ahead we could see the sinuous line of an advancing bush fire: a serpentine of smoke enclosing flickering orange glints of flame, behind which lay a desolate plain of smoldering gray ash. Casually, browsing as they went, the animals walked ahead of the fire, with many species keeping good company. There was no *Bambi*-style panic here, for the winds were mild and the fire slow-moving.

Finally we turned north, looking for the kopjes of the Chobe. In particular we wanted to find the Shinamba Hills, which were indicated on our maps with a question mark. We wouldn't have been able to precisely locate them if we did spot them, because here there were no landmarks to take bearings on, but at least it would have been nice to know for sure that they really existed and ascertain their approximate location. Don didn't think that he had ever seen them, even though he had flown the region often enough. In spite of making a number of traverses across the area in which they were supposed to be, we couldn't find them, so I signaled Don to return to Maun. We would have to find the Shinamba Hills on the ground.

Dave led the way to our camp, where he had spent the morning supervising the crew in getting set up and provisioned. It was located several miles north of the town in an isolated patch of bush, though not so far from the river that we couldn't get a daily supply of fresh water. Deacon, our campkeeper, was there to greet us when we arrived. With a fire going and pot already aboil, he offered us a cup of tea. With cup in hand, we inspected the setup that would be our home for the next few months. The camp that Teddy and I would occupy consisted of two sizable rectangular canvas tents, one for sleeping and one for stores and equipment, facing each other across a tarpaulin-carpeted area furnished with a large steel table and chairs. The sleeping tent was furnished with steel cots hung with mosquito netting and with footlockers for our personal gear. Dave had temporary quarters in the equipment tent, since he would be with us only for a week to help get us started. Deacon's kitchen was adjacent to our living area, and consisted of a cooking fire laid out on the sand, a table for preparing the food and washing up, and his collection of cooking utensils and washtub. The crew's camp, four circular tents with benches, stools, and its own cooking area, was twenty or so yards away, behind Deacon's work area and screened off from our camp by a few trees and scrub bushes.

Around our camp were strung wires from which hung 12-volt DC light bulbs that ran off either batteries or the little gas-driven generator we had brought along for charging our instrument batteries. They produced a pretty feeble yellow glow, though, so we ended up using the Coleman lanterns we had fortunately thought to bring with us from the States. The bath was a collapsible canvas bucket fitted with a shower head. It was suspended from a tree over an enormous galvanized bathtub, and activated with a rope. We had as much gear as Boot, the reporter in *Scoop*, except for the folding canoe and the forked sticks for sending messages. Teddy went around saying, in his bogus British accent, "Splendid! Simply splendid, old chap!"

For the pièce de résistance, in the stores tent I discovered a propane fridge, which contained a nice-looking piece of steak and a six-pack of Castle's. "Cold beer! All right, Deacon!" I held the beer up in toast to Deacon, who had been anxiously watching us make our inspection, and who grinned in response. "Reeb, Ted? Dave?"

"I thought you'd never ask." "Sure, I'll have one, too."

We sat down at the table and started working on our maps while Deacon began preparing dinner. In working out our plan of action we had two big problems: no good maps and no good rock. The best maps available were some British military maps at five miles to the inch. These "Joint Operations Graphic Ground" maps had evidently been made from satellite images with very little ground control. They were covered with stamped provisos like "THE USER IS WARNED THAT THIS DIAGRAM IS NOT BASED ON ACCURATE DATA AND MUST BE USED WITH EXTREME CAUTION" and "OWING TO INADEQUATE DATA ON EDITION 1-GSGS OF THE ADJOINING GRAPHIC THIS EDGE WILL NOT MATCH." There were confidently labeled place names like "Shinamba Hills?" We weren't sure if the question mark referred to the name, the location, or the existence of the indicated hills. Our flight hadn't helped out too much on that one. The main and only road into the Chobe National Park from the south was enigmatically marked "Route closed 1964." Was it later reopened? Was there another route now? This lack of precision in our maps posed a serious problem for laying out, even provisionally, our seismic networks, because in planning sites for the instruments it is kind of important to know if you can get there by some reasonable means, and, once there, to know where you are.

The rock problem was serious for a different reason. The Geological Map of Botswana is one of the more boring I have ever seen. Except for a region along the southeastern edge of the country, it is almost everywhere colored a uniform yellow. According to the legend, this signified "Kalahari Beds," meaning sand. If you want to listen for tiny earthquakes occurring at five or ten kilometers' depth in the crust, it will not help at all if there is a blanket of a few hundred meters of sound-absorbing sand covering everything. Recording on sand would allow us to detect microearthquakes only down to magnitude 2, rather than magnitude 1, which we could detect with instruments sited on rock; in consequence we would be able to detect far fewer events. Trying to pick up earthquakes through a layer of sand was like trying to eavesdrop on your neighbors by listening through a pillow propped up against the wall. Unless we could get rock sites we would be severely limited in the number of earthquakes we could expect to observe in the time we had allotted ourselves. This meant that our plan had to revolve rather heavily on getting instruments on the few kopjes in the area, which were widely spaced and in very remote places, mostly with no marked access by road or track. At least on our flight we had found the southern kopjes, though the location of the northern ones was still in question.

We had no such interesting difficulties with our initial setup, though. Since there is no rock anywhere near Maun, one site was just as good as another. So it didn't take us too long to lay out a network that was geometrically OK and looked as if it had reasonably good access. We had just agreed on a work plan for the next day when Deacon brought on the dinner. Dave dove into his with relish, but Teddy and I, after a few nibbles, poked the food about our plates, longing wistfully for a dog. The steak was cooked to dry shreds, the potatoes were overboiled and underseasoned, and the canned peas were, well, canned peas. Teddy gave me a look, and grumbled, "Awfully British, don't you think, old chap?" "Awfully," I replied. "Methinks that Deacon starts remedial cooking lessons on the morrow."

That evening as we sat around the table having a drink before turning in, the sound of drumming started up, distantly reverberating from out of the darkness beyond the circle of lantern-light. Teddy looked at me, wide-eyed, and said, "Do you hear what I hear?"

"Yeah."

"Holy shit, what have you got me into now, Scholz?"

Thomas showed up promptly at eight the next morning and we loaded up two Land Rovers with our gear, with a driver and two helpers assigned to each. I sent Dave off with Teddy to get trained on the equipment, while I went off with Thomas to start my own education on the ways of this land. I directed Thomas first to cross the river and drive along the low sand hill that I believed to be a fault scarp. I thought that this was the most likely place to find rock anywhere near the surface and that if we saw any sign of it, we might find basement outcroppings if we dug down a bit. After driving at a crawl for more than two hours, I hadn't been able find any place that looked any more promising than any other, so I called a halt and asked the laborers to dig a hole at a spot on the crown of the hill. While they set about this chore, I wandered off, looking for "float"—bits of rock in the sand that might indicate outcrop at depth. Of this I found not a trace; everything was covered by the same monotonous blanket of sand.

On returning fifteen or twenty minutes later, I found Thomas sitting in the shade of a tree. In front of him was a big pile of sand next to a hole from which the tip of a shovel emerged periodically, flipping sand onto the pile. Whatever other talents these guys had, they were certainly prodigious diggers. I walked over and peered into the hole. The men were already seven or eight feet down and showing no sign of letting up. "Whoa, hold it," I said. "That's deep enough." They had made a square hole in the sandy soil, three feet on a side, eight feet deep, with perfectly vertical walls. If they dug any deeper there was a good chance of it collapsing in on them. They clambered out with the help of a pole, and I lowered myself down. Examining the sides of the hole I could find no signs of layering, calcrete or otherwise. It was just uniform sand, becoming more compacted and cohesive with depth.

I had them lower the geophone down to me, which I installed in the bottom of the hole, placing over it one of the pyramid-shaped wooden boxes to isolate the geophone that we had had made up in Lobatse for this purpose. I climbed out, unreeling the electrical cable behind me, and instructed the crew to descend and backfill the hole while carefully tamping the sand around the geophone box. When they were through, I had them cut down all the trees for ten meters around the site. Another problem with sand was that, being very compliant, it picked up any wind vibrations transmitted into the ground by trees swaying in the breeze, so we had to clear the areas around the instrument sites. The men were as proficient at chopping as they were at digging, and in a few minutes the job was done. I set up the recorder under the nearest remaining tree, connecting it to the two car batteries that were the power source and plugged in the geophone cable. The results were not encouraging. The highest magnification at which I could set the amplifier was 60 dB before I encountered excessive ground noise. This was sixty to one hundred times less sensitive than an instrument installed on bedrock. It was bad news for detecting many microearthquakes.

My next concern was about instrument security, but when I asked him about it, Thomas didn't think we had to take any special precautions about leaving the instruments in the field. There weren't any large animals in that area, and he didn't think the people would bother them. In this he proved correct, which was a big relief. In "civilized" places like California I've come back to find bullet holes in recorders. In northern Baja California I even had the Federales walk off with one once, under the impression that it was some kind of Yanqui spy gadget. That one disappeared somewhere into the Mexican military organization, and I never did get it back. Here all I had to do was place the recording unit where it would be in the shade all day. It was already 100° in the shade, and the stainless steel box that held the recorder wouldn't have to be in direct sunlight too long before some part of the electronics would get overheated and go on the fritz.

We installed two other instruments that day, with similarly disappointing results. I got only slightly better results at one site, where we found a calcrete layer, a natural caliche deposit, about six feet down, and managed to record at 66 dB. Derek was right about calcrete being better than sand, but it still wasn't anywhere near the realm of good recording conditions. So I was a tired, hot, dusty, and not very happy puppy when we arrived back at camp a little after six.

I was just opening a beer when Teddy and Dave pulled up. They looked about as fried as I felt. "How'd it go, you guys? You look like you could do with a reeb."

"Sure, Chris, I think I could choke one down. Toss one over."

They had much the same experience as mine to report. When I told them about the calcrete bed, Teddy said, "The hell with digging for calcrete. I found a much easier way to get 66 dB. Termite mounds. Heh-heh, no shit. Just stick 'em in termite mounds. You get 66 dB every time."

For the next two weeks we toiled mightily on this problem, but did not find any solution. The only thing we discovered was a flaw in Teddy's anthill theory. You had to put the geophone in the inactive part of the anthill. Once we put one in the active part and on returning the next day found it tipped over and cemented in. How the little buggers managed to tip over a five-pound geophone, I'll never know. So we were stuck with low-gain recording. This was very frustrating, because every day or two we would record a tiny "event" but it was usually too small to be recorded on more than one station, so that we couldn't locate it. Earthquakes are located by a kind of triangulation procedure, so they have to be recorded by at least three seismometers in different locations before that is possible. Unless we could locate the earthquakes, we couldn't say anything about their pattern of activity: whether or not they were associated with our suspected faults and if they showed a sense of movement consistent with my hypothesis about how the faults moved. Recording small events on a single station was tantalizing because it told us that the region was in fact tectonically active. If only we could hear them better, we could find out what they had to tell us about what the Earth was doing. "Listen to the Earth, and it shall teach thee," is the motto of the seismologist, paraphrased from the Book of Job.

In the two weeks we worked around Maun we were able to record only a half-dozen events large enough to locate. I determined rough locations for these with my pocket calculator. They all lined up a few kilometers to the west of the Thamalakane "fault."

This was in agreement with my hypothesis, because if that was the fault bordering the eastern flank of a rift valley it should be inclined steeply to the west, and any microearthquakes occurring at depth on that fault should lie to the west of where it outcrops the surface. This was all fine and dandy, but it sure was slim pickings. Mother Nature was being more reticent than usual in revealing her secrets. Unless we could get a lot more data than this we would have a hard time making a convincing story out of it.

Those two weeks were well spent, though, as a shakedown period in getting ourselves organized in the work and knitting ourselves into an efficient field operation. The crew quickly learned their duties and responsibilities, some of them completely new to them, under the guidance of Blackie and also of Thomas, whose local knowledge was invaluable. We were still close enough to town that the various bugs and deficiencies that turned up could be quickly remedied. In this Dave Potten at the U.N.D.P. was especially helpful. It was reassuring to know that there was someone in that otherwise somnolent organization willing to react with imagination and dispatch to any of our problems and requests. We quickly learned to deal with him exclusively, and not to trouble Dr. Ernest, who was not much help in practical matters and would, after dilly-dallying, refer us eventually to Dave, anyway. On our many trips to the office we got to know the two scientists in the Maun office, an agronomist from Belgium and the Swedish hydrologist. The Swede and I chose not to get along. He, rather snottily, I thought, expressed the opinion that studying earthquakes in the Okavango was an irrelevant waste of time and money. I, for that matter, was unable to convince him that all his meticulous stream-flow studies would be of little use if a sizable earthquake came along and put a tilt in the whole drainage system. To each his own.

Meanwhile, we learned to pace our work in a way more sensible for the climate, taking lengthy breaks during the midday for a light lunch and siesta in the shade of a tree. Between five and six in the afternoon we would meet at a place a few miles north of the town where a plank bridge crossed the river. There the river flowed, with a smooth and surprisingly swift current, between reedy banks in the shade of a tall grove of trees. Like all the water in the swamps, the river water, filtered through the clean sands and devoid of scarcely any upstream human habitation, was potable and clear, and free of bilharzia and the other equally unpleasant parasites that are endemic to so much of Africa. In that idyllic spot we would swim for half an hour, and perhaps do a bit of bream fishing, while our crews, knowing neither how to swim nor fish and being uneasy around so much open water, took one of the vehicles back to camp. When later we would arrive back in camp we would be thus clean and refreshed, ready to enjoy a few beers before dinner.

We were engaged in this pleasant activity one afternoon when a pickup stopped on the bridge and the driver shouted a greeting to Dave. It was some people Dave knew from Lobatse, whom Dave introduced to us. The driver's name was Joe Oppendorp, a big heavy-set Afrikaner in his mid to late fifties. His pal, Clyde, whose last name I don't recall ever learning, was of a similar age but slightly built and balding. Joe greeted us ebulliently, and invited us to their camp for gins and tonics. Such a luxury could scarcely be refused, so we quickly dressed and, piling into our Land Rover, followed them to their camp.

They had picked a beautiful campsite, on a grassy bank overlooking the river. They had a big pickup equipped with an oversize camper with all the modern conveniences. In front was pitched a spiffy net-sided pavilion tent, set up with folding lawn furniture, which served as their living room. After proudly showing off his camper rig to us, Joe ushered us into the tent. He bade us sit down and get comfortable while Clyde busied himself in the camper fixing the drinks and snacks, which he brought out on a big tray. Joe was effusive and jolly, in a hail-fellow-well-met sort of way. Clyde, on the other hand, was quiet and diffident, showing a kind of deference to Joe that suggested the servant, or perhaps something else as well.

The drinks were very welcome, and the conversation boisterous and jolly, but there was something about Joe's manner, or perhaps it was the situation, that put me a bit on edge. I was reminded somehow of the time I was five and the Jehovah's Witnesses who lived down by the river in Gold Hill, Oregon, coaxed my sister and me into their converted schoolbus on the promise of grape juice and cookies. They were such smiling, friendly looking people that it didn't seem proper to refuse, and besides, we wanted the grapejuice and cookies. Once inside, I was very uneasy. There was something off-key going on, even a five-year-old could feel it. When they gave us the religious pictures and started telling us stories about them in slightly weird voices we knew it was time to get out.

On the way back to camp, Dave filled us in about Joe. He was a labor contractor for the South African mines. His job was to travel around the southern African countries, visiting villages and talking the young men into signing three-year contracts to work in the mines, and then shipping them out. The carrot was that with the savings from three years' wages they would be able to return to their villages with enough money to buy a wife and a plot of land to start their lives with. The problem was that after three years of grueling servitude underground, living in barracks and off the company store, things seldom worked out that way. So Joe was more or less in the same line of work as the Jehovah's Witnesses.

On his last night before returning to Lobatse, we took Dave to Riley's for a meal and piss-up. We made quite a night of it, meeting Joe and some of his pals in the bar during the evening. Clyde wasn't with him. As Joe said, "Clyde doesn't go in for the pub crawlin'. Enjoys his quiet at the camp, he does."

We had quite a few rounds of darts accompanied by that many more of beer. Sometime during the evening Dave focused on me and said, "By the way, what does `reeb' mean?"

"It's what you're drinking, Dave."

"I sort of guessed that. What I meant was, why do you call it reeb?"

"Why do we call it that? I think it's because in American bars they always advertise it with a neon sign in the window that says `reeb.' "

"Actually," piped in Teddy, "it usually says `reeb dloc,' but that's too hard to say."

7

There is no town or village at Toteng. It is simply a locality, a place with no more merit than a name to indicate a particular spot, and useful for filling in blank spaces on maps of such cartographically empty regions as the Kalahari. We set up our camp on a low bluff above the Nhabe River, which drains the Okavango down from the north into its saline sump, Lake Ngami. There were a few huts on the other bank of the river; otherwise there was nothing there of note but a low bridge of wooden poles, newly reinforced by I-beams stenciled "U.S. Steel," on which the Maun-Ghanzi road crossed the river. The campsite was ideally situated. There was a grove of trees to shelter our tents from the intense sun, the river provided plenty of clean, fresh water and a welcome, schistosome-free bathing hole, and we had easy access to the whole of the surrounding regions.

Having shifted our base from Maun, we were mainly working to the south of Toteng in the basin of Lake Ngami. The lake level was very low at that time. The other primary inlets to the lake, from the southern regions of the Okavango, had been blocked off by heavy papyrus growth in previous years, or so it was theorized. That part of the swamps hadn't been surveyed for years, and since almost no one ventured into that area, nothing very sure was known about it. The edge of the Lake Ngami basin was covered in a veld of thorn trees and scrub, giving way in the interior to a barren region of dry lake beds and salt flats. Once one entered the flats, the only signs of life were occasional crowds of vultures and marabou storks, stripping a carcass of a cow or some other animal that had wandered into that thanatosic plain. Far across the lake, in the shiver of mirage, we could just make out the flocks of flamingos and ibis that feasted on brine shrimp in the shallows. It was a forbidding place. I could not say that this part of the Kalahari evoked in me Bowles's Saharan image of a "puritanical Eden."

The basin focused the heat like a lens, and Teddy and I were beginning to suffer from it. This was the first time we had been exposed to the full brunt of the Kalahari sun. The thermometer in camp would register 120 degrees in the shade by midday, but by an hour out of camp we had left any shade far behind. We took to hitting the water bags at every stop and glugging it down. Ours was no ordinary sense of thirst—it was like the cravings of an addict. It could be appeased for a few minutes, but it wasn't for much longer that a drink of water wasn't again foremost on the mind. This incessant thinking about water, like the tug of nicotine deprivation, lasted all day long—it didn't subside until an hour past sunset, when we had had a bath in the river and a few beers. We were drinking gallons of water a day, and yet our skin was never the slightest bit moist. The crew, of course, didn't share our affliction. They would have a good drink at the midday meal, and that was about all. It was us who were always calling for halts as an excuse for a shot of water. This was a little embarrassing, and we struggled mightily to extend the time, bit by bit, between drinks of water, with each additional minute being a conscious struggle. It would be many weeks before our systems would acclimatize to these conditions and we could get through the day without this constant craving for water.

For the first few days we were busy setting up the main body of our portable seismic network. This required a lot of reconnoitering in the Land Rovers, since only the main road was marked on any map, and that only schematically. It was as though the mapmaker had been aware that there ought to be a road going from Toteng south into the central Kalahari, but because its exact position changed from year to year there was no point in being too precise about where to draw it. All the other tracks, and there were many, were no more than where someone had driven since the last rainy season. We had to make up our own sketch maps to find our way, and we also had to locate ourselves on our base maps as best we could, since we had to know the relative locations of our seismic receiving stations as precisely as possible. This required a lot of guesswork and tricky calculations, since there were scarcely any topographic features, and in those days there were no satellite navigation systems like GPS that could make locating oneself in remote areas easy and reliable. But we had no choice. If we were going to be able to locate the microearthquakes we were hoping to detect, we had to know the whereabouts of our recording stations.

After we became familiar with the area, we split up, each taking a Land Rover and crew out on our various tasks and rounds, in the same routine we had established in the initial surveys around Maun. We were still bedeviled by bad siting characteristics, but in this district we had three possibilities to find bedrock sites for the instruments. The first was a spot locality of a rock outcrop, indicated with a star on the Geologic Map of Botswana as being somewhere in the vicinity of Toteng. The second possibility was to the east of Lake Ngami, where we had been told a U.S. Steel Company geologic field crew was doing exploration work. Perhaps they had opened up test pits to bedrock. The other was the Kgwebe Hills, which we had already checked out on our recon flight.

It took a couple of days to find the Toteng site. The star marking it on the 1: 2,000,000-scale map covered a couple of square miles of scrubland, and since the outcrop itself was only a small nose of rock sticking out of the sand, it was only marginally easier to find than the proverbial needle in the haystack. On the day Teddy found it, I was down at the south end of Lake Ngami looking for the bounding scarp, that will-o'-the-wisp that was so obvious on the satellite images but so hard to see on the ground. I was ostensibly looking for it because it offered the chance of finding bedrock uplifted and close to the surface. After my experiences of traipsing along the same feature a hundred kilometers to the north along the bank of the Thamalakane, I actually had little hope of finding rock, but I wanted to see the scarp anyway. After searching for the most of two days, I finally convinced myself that I had found it. It appeared as a low, one-sided hill, or monocline, just above the uppermost of the old lake shorelines, now high and dry. It was about ten meters high, and covered with a calcareous flinty scree. The shorelines, defining a set of shallow terraces descending to the level of the present lake, were congruent with it, showing that it was the master template defining the architecture of the basin as a whole. Here, for the first time on the ground, I could see my vision come alive before me. In my mind's eye I could see this fault scarp growing, in sudden jerks, each time an earthquake occurred on the fault. Gradually the Okavango would be dammed behind this growing scarp, empounding the swamps upstream and here forming this lacustrine basin. For the first time, I unequivocally recognized our target. To bag this prey, which meant to identify and understand this process, was the purpose of our search for microearthquakes, since they would provide the crucial, decisive evidence, the kind of evidence that would hold up in the court of science. Collecting this confirmatory evidence was ordinary scientific work, the job at hand. As for myself, I already had little doubt, as I stood on this small ridge in this nearly featureless plain, that I was standing at the tip of a future African Rift Valley. If all things worked out, I would later be able to show that at this point is the very tip of the rift system that extends all the way from the Afar triangle, in Eritrea and Djibouti, where the East African rifts are born at a triple junction with the oceanic rifts in the Red Sea and the Gulf of Aden. This great rift system has been propagating southwards through Africa for 30 million years, but had just recently reached this place in the Kalahari.

We met Teddy and his crew lounging under a big tree on the road back to Toteng. As we pulled up Teddy announced, with a big grin, "Hey, we found Toteng Star."

"Phou, what a relief. I was beginning to wonder if it really existed."

"It ain't all that big. You won't believe it. You could have driven right over it in the dark. We wouldn't have found it except for the cows."

"The cows?"

"Yeah, there were cows standing all around it. It must be salty, or something. Anyway, it's 120 dB."

"A hundred and twenty dB!"

"No shit. Heh, heh. Come on, check it out."

We followed their Land Rover along a track that threaded its way through a maze of scrub, finally coming out upon a barren flat in which stood a desultory looking herd of cattle, beyond which was that rarest of sights in this desert—a low outcropping of rock. We got out and walked around it. I was beside myself with glee, while Teddy looked on like a proud papa. No more than two feet high and ten feet across, it was a piece of honest-to-goodness Precambrian basement poking its lovely, sensitive nose out of the desert floor. I hit it with my hammer and it rang like a bell. This was no loose boulder. It was a solid outcrop of peridotite, a chunk of 2 billion-year-old mantle that was directly connected to where the action was.

Teddy had set up the instrument on the far side of the outcrop. He had carefully sheltered it from the wind, though there wasn't a breath of it in the heat of midday, and had anchored the three legs of the seismometer firmly to the rock. I checked all the instrument settings. There was no doubt about it, it was running at 120 dB gain, a magnification of three million, and without a bit of background noise. The amplifier was running at maximum gain, something I had never seen before. I thumped lightly on the rock with my foot, and big sharp signals appeared on the recording drum. This site was about six hundred times more sensitive than any instrument we could run in the sand. If there was any microearthquake activity at all in this vicinity, we would soon know about it. We finally had a big ear to the ground.

These suspicions were confirmed when we checked the Toteng Star instrument the next day. It had recorded three microearthquakes overnight, each within twenty-five kilometers of the station. By contrast, the other instruments sited in sand had recorded only one such event in the previous five days, and that one had been far to the north, near Maun. Though there was now no doubt that the Ngami Basin was seismically active, this information was as tantalizing as it was gratifying. To locate these tiny earthquakes, we had to get three high-gain instruments, and it was obvious that the ones installed in sand were not going to do the job. If we had plenty of time to wait, eventually larger earthquakes would occur that the low magnification instruments would register. But we could spare only about ten more days at Toteng before we had to move on to the north. We were on a tight schedule to ensure that our whole survey would be completed before the rains came and made all roads impassable. This made it essential to occupy the other rock sites that we knew existed, and as soon as possible. Every day they weren't operating meant another day of lost data, and we had only so many days left.

That evening Teddy and I went over our options. We knew full well where the Kgwebe Hills were. They were a group of kopjes, rocky hills that at irregular and distant intervals poke out of the Kalahari, remnants of the old Precambrian basement of which Toteng Star was but a taste. The Kgwebe Hills were a group of four or five of these, five hundred to six hundred feet high and arranged in a rough circle. Not only were these the only kopjes in a hundred miles and clearly marked on the maps, but we had flown low over them on our recon flight at the start of the fieldwork. They were less than fifty miles away, as the crow flies. The problem was that we didn't know how to get there, and our trusty guide Thomas was not being very forthcoming in that regard.

"Ted, we've got a serious problem. We have to get to the Kgwebe Hills, but something's fishy about that whole subject. Ever since we moved down here to Toteng I've been bugging Thomas about going there, but all he ever comes up with is something like 'you can't get thar from here,' like he was some kind of Maine potato farmer."

"Well, maybe you can't. We haven't found any roads going east."

"I don't believe it. In this kind of country there are always tracks going anywhere of the slightest interest, and that has to include the Kgwebe Hills. You saw for yourself how green they are. There's water there, which is about as interesting as you can get around here. Besides, when we flew it we saw tracks coming in from both the west and east. "

"Yeah, I know. You're right about there being something funny going on. I get the feeling that the crew doesn't want to go there. I was talking about it to Straing yesterday. You know, he's the young guy I've been bringing along. He didn't exactly come right out and say so, but I got the impression that the crew has been talking about little else but the Kgwebe Hills since we got here. It seems like they're scared to death of going near the kopjes because they are supposed to be Bushman holy places, or something, and Thomas has been filling them with all kinds of stories about the Kgwebe Hills."

"Ah, so that's it. I was wondering if it might be something like that. Stories about the Bushmen and the kopjes are famous legends in this part of the world. Van der Post wrote about it in *The Lost World of the Kalahari*. According to him, the Bushmen believe their Gods live in the kopjes, and the Tswana, who are scared of the wild Bushmen anyway, think the Bushman spirits haunt them."

"Oh, great. First we got no outcrops, now we got haunted ones."

"It makes sense with the way Thomas has been acting. Every time I bring up the Kgwebe Hills he gets real chuffy and evasive. I get the distinct impression that he knows how to get there but just doesn't want to let on. He hasta know, he's lived around here all his life, for Christ's sake."

"Fine, but if he won't tell us, how do we get there? Seems like a Mexican standoff to me."

"First, we have to find that U.S. Steel camp. They're sure to know the way, and they aren't likely to be scared off by any Bushman ghost stories, either. After we get proper directions, we can damn well go by ourselves, if the crew doesn't like it."

"Hmm. Fair enough. By the way, what are these Bushmen kopje stories? Just curious, heh-heh."

"Oh, come on, Teddy. It's just a lot of hokum. Van der Post told this story about when they were filming 'The Lost World' and they couldn't film the old Bushman petroglyphs in the Tsodilo Hills, supposedly because they were protected by a nest of cobras. Then they got chased out of the kopjes by a swarm of bees. He tried to make out that these were some kind of Bushman spirits or something. Hey, what the hell, it added some zip to his book. You've gotta sell copies somehow. I've got a better story than that one. Did I ever tell you about what happened when Bob Smith and I put an instrument on a sacred Navaho place, Leche Rock, out of Page, Arizona?"

"Okay, I'll bite, heh, heh. What happened?"

"We got chased off by a Navaho with a rifle. Some spirit, huh? Spirit of the Old West."

We headed out of camp the next day in caravan with both Land Rovers and crews. The plan was to follow the main road all the way to Ghanzi, looking for any well-trafficked route to the east. When we were in Lobatse I had been told by people in the Botswana Geological Survey that the U.S. Steel camp was somewhere to the east of us, between Toteng and the Kgwebe Hills, but so far the only signs we had seen of their presence were the new I-beams in the Toteng bridge. I was curious to visit Ghanzi, anyway, which was an extremely isolated old Boer farming community in the central Kalahari that dated from pre-protectorate times. The Kalahari Arms, the pub there, was if anything more notorious than Riley's in Maun, and was the last stop before Windhoek, in Namibia, a long, long drive across the desert. But as it turned out, our business didn't take us that far, and we never did get to the Kalahari Arms.

As we descended the long gradual slope into the central Kalahari, we came upon a strange sight. A proper-looking steel road-sign, unheard of even in the populated parts of Botswana, stood next to the dirt track in a landscape in which no other feature could be seen in any direction. It said, with an arrow pointing forward, "Ghanzi," and, with an arrow pointing to the left, "Ngamiland Playboy Club."

We stopped and took a picture of the sign. Teddy said, with a grin, "I guess we found the way," and off we went down the road to the left. After another hour we came upon the camp in a grove of trees. It consisted of several long, low buildings, like bunkhouses, a lot of sheds with heavy earthmoving equipment, some small cottages and office sheds, and one larger building in the center, where we pulled up. Teddy and I knocked on the door and then peered in. It was a dining hall and clubhouse, with tables in the middle, a bar and dartboards to the left, and a kitchen off to the right. There were half a dozen white men at a table, looking at us with startled expressions.

"Hi. Is this the U.S. Steel camp?"

A giant with a blond beard and a South African accent said, "That it be. I'm Bill Mackay, Exploration Manager. And who might you be?"

We had fortunately arrived at lunchtime, which was not only a good time to catch everyone in from the field, but also to get some lunch, which was steak and chips, a far cry from the sort of grub we usually packed for a midday meal. Like any group of men who spend long periods together in isolation, these geologists and mining engineers treated us, their rare visitors, with a mixture of warmth and guardedness. Yet professional courtesy prevailed, and eventually, after ferreting out what our business was over lunch, they opened up and were quite free, at least as far as exploration types go, in telling us what they were up to out there. (Exploration people are notoriously secretive, so I knew right away that it must not be that hot a prospect, otherwise they would have just sat there like so many clams.) They were working on a large sedimentary iron ore deposit in the Karoo, which is the rock underlying the Ngami basin. They had been out there for going on three years, evaluating the tonnage and grade of the deposit and planning mining schemes, although they had long ago reached their own conclusion that it wasn't likely the deposit would be developed anytime soon. Mackay told us, "It would be a great deposit if it were somewhere else. Here, there's no way to get enough water to mine it, and it's just too far from the railroad. It's big all right, make no mistake, but it would have to be bloidy huge to pay for enough infrastructure development to make an economic go of it."

We eventually got around to our particular problem. "Yip, we got some exploration trenches 'dozed down to rock, though I wouldn't exactly call it bedrock—it's in the mineralized zone. Take the north road out of camp. You'll see them a few miles up on your left. Can't miss 'em. Nope, nope. No problem, go ahead and put in your geophones. We're not working out there just now, so we won't bother them."

"The Kgwebe Hills, eh? No problem that I know of. Just keep going past the trenches. You'll see the way. What's the Kgwebe road like, Fred? Hmm. Well, none of us have actually been up there, but it shouldn't be any problem. What transport have you got? Land Rovers? Nah, it shouldn't be any problem at all."

I had our geological map of Ngamiland laid out on the table during these discussions. "By the way," I said, indicating the map, "have you guys got any better maps or sections of the basin that we could have a look at?"

"Sure, we can show you a few things. Not too much of our stuff is classified, anyway. Come on over to the office."

We walked across the road to one of the office sheds. Inside was a big open room filled with worktables. Along the back wall was a rack containing drilling cores and maps rolled up in tubes. Bill pulled a few of these out and unrolled the contents onto one of the tables. He spread out a big geological map of the Ngami Basin and a set of geological cross-sections. They were hand-drafted in india ink and colored pencil, and full of details obtained from boreholes. The map was dominated by two faults, on either side of the present Lake Ngami, striking north-northeast across the basin. "Wow, look at those faults! Those aren't the presently active ones, are they?"

"No, no. Those are Paleozoic in age. They cut both basement and the Karoo sediments, but as you can see from the section, they don't offset the sediments and the youngest Karoo member." I studied the cross section. On the east-west section, the faults were mapped as normal faults, with the Karoo sediments in the middle being down-dropped to form a shallow graben. "This is one of the late Karoo troughs," he went on. " Many of the Karoo basins were rifted just before the eruption of the flood basalts."

"But there are no basalts here, right?"

"Oh, no. There haven't been any Karoo basalts found this far to the northwest. They are mainly confined to South Africa and Mozambique."

"Hmm. Well, this is very interesting. The idea that we're following is that there is a modern rift basin just starting here, with a northeast-southwest trend. It's interesting that it almost coincides with this old one. I'm wondering if there is a connection. Thanks very much for letting us have a look at these."

"No problem. Say, if you fellows are going to be in the area for a while, drop in some evening. We'll give you a few beers and a game of darts. We don't get much in the way of company out here."

The rock in the trenches turned out to be pretty broken up and crumbly, nothing compared to the rock quality at Toteng Star. But it was still a far better site than any on sand, so we could honestly claim that we had our second rock site installed. From the trenches we could just see the Kgwebe Hills looming on the horizon to the northeast, but since it was late in the day I decided to postpone that visit until the morning. But I pointed out the road and told Thomas my plans.

"No, boss. This very bad road. We can't go to Kgwebe Hills this way."

"Do you know another way?"

"This is only way. But road washed out. It break Land Rover."

"Well, Thomas, the geologists we just talked to at the U.S. Steel camp told me that this road was fine. Anyway, I don't really care what the road is like. We're going to try it tomorrow, washed out or not. We'll bring the pickup with the winch along just in case we get a vehicle stuck. Tomorrow we're going to get to the Kgwebe Hills even if we have to break new track to do it."

Nothing more was said on the matter until that evening. After dinner, Thomas came into our camp. This was something he had never done before: It was a breach of camp etiquette. Normally the crew stayed in their own camp during off hours.

He stood in the shadows a few feet from our mess table. "We can't go to Kgwebe Hills, boss. It's too dangerous."

"What do you mean, dangerous, Thomas?"

"Kgwebe Hills is Bushman place, boss. Too dangerous for us to visit. Da crew, dey very scared of that place."

"Well, maybe they are, Thomas. Or maybe they've been listening to too many stories. But we are going to the Kgwebe Hills tomorrow. If the crew is afraid to come, fine, they can stay in camp. Teddy and I will go by ourselves. We know the way."

I can't say that I was completely without any qualms at that point. I was not concerned with the superstitions of the crew, but I had grown accustomed to following Thomas's advice regarding bush-lore, and I didn't feel too comfortable in disregarding it this time, no matter how ridiculous it seemed to be. If there was a real problem, he wasn't exactly being clear about what it was. I certainly had no intention of backing down in the face of such a vaguely stated danger. Not only was a site in the Kgwebe Hills essential to our network, but there was also the matter of maintaining my authority with the crew. It doesn't take too much backing down before you end up getting a total runaround from a local crew. I tried to remember exactly the reaction of the U.S. Steel guys when I had mentioned the Kgwebe Hills. It seemed to me that they had shown no particular reaction. Still, it was curious that none of them had ever visited those kopjes. Although their iron deposit was of Paleozoic age, and so was much younger and geologically unrelated to the Kgwebe Hills, I should have thought that any geologist stuck out here for a couple of years would at some point, if only out of curiosity, visit the only rock outcrop in the entire district.

At eight the next morning the Land Rovers pulled up to our camp, with Thomas and Samuel at the wheels and the crews in the back, just like any other day. We loaded up the usual gear, including this time the special instrument with the double recording drum that we reserved for particularly remote sites. As we rolled out of camp, I noticed that the Ford pickup was not coming along, after all. So much for the washed-out road story.

Our work that morning proceeded normally. When we stopped to check the instrument at the exploration trench, we found that it had recorded an event during the night, confirming our expectations of good hunting.

The road that continued on to the Kgwebe Hills was a substantial one. It had been graded at one time and was still in pretty good shape, although it had obviously not been used for a long time. The hills themselves appeared as gray lumps on the horizon. As we approached them, Thomas and the crew, who normally chattered throughout the day, fell silent. We continued to ascend the gentle alluvial fan, and the kopjes seemed to rise out of the desert floor like immense, shiny gray dumplings. Baobab trees and leafier, more succulent desert shrubs began to appear, a sure sign of rock and a water table at shallow depth. Suddenly there was rock on both sides as the road entered a cleft filled by a dry stream bed. We began to climb. After a few minutes the narrow canyon opened into a valley with a stream bed, in which there were, here and there, stagnant pools of water. Grass grew along the banks, which were surrounded by mopani and willow trees.

These ecological zones, first through the baobabs, then in this lush stream valley, were so different from the dry veld we had been traveling through for so many weeks, and the changes so abrupt, that they presented a very strange aspect to our minds. It was as though we had been suddenly transported into a different world. Even the light was different here, illuminating everything with a gentler but more penetrating glow than in the surrounding desert, without the dusty haze. Here we were hemmed in by rocky cliffs: For the first time in many weeks the land wasn't a flat disc extending to the horizon in all directions.

Soon an even more startling sight appeared. It was a house, an old Dutch stone farmhouse of the same eighteenth-century design as the ones at home in the Hudson Valley, sitting in a beautiful setting in a grove of willows. What in the hell was this doing out here in the middle of nowhere? We hadn't seen a stone house since we left Francistown; even mud huts were extremely rare in this region. But there was something else odd about this house, which became clear as we got nearer.

"Uh-oh."

It was completely burnt out, and didn't look as if it had been occupied in decades, maybe even centuries. "What happened here?" I thought. "There must be some story behind this." The Afrikaner voortrekkers who settled this land, especially at the time when this house was built, were not the sort of people to be easily chased off their land. Especially not land as fertile and well-watered as this. Just thinking about it was enough to give you a case of the willies.

We passed the farmhouse and climbed more steeply as the valley narrowed. It was now more heavily wooded, and water was flowing in the stream. The low murmuring from the back of the Land Rover had ceased and a deeper silence now prevailed. I glanced over my shoulder. "Huh?" Talk about the willies. The laborers were huddled up fetally on the side benches and the floor with their arms over their heads. I turned around and watched the road, without a comment to Thomas, who drove intently and silently on.

At the top of the incline the road suddenly spilled out into a valley, where we came to a halt. It was the same beautiful valley that we had seen from the air a month earlier. It was like a Shangri-la, a circular, flat-bottomed valley, maybe a mile across, surrounded by the kopjes on three sides, and green with vegetation. On the far side of the valley there were leafy green trees, like banana trees. We could also just make out what might be the brown thatch of hut roofs here and there among the trees, and what looked, oddly enough in the circumstances, as if it might be cultivated fields. There were no other signs of habitation, and no one to be seen on the side of the valley where the road emerged.

Teddy, Thomas, and I descended from the Land Rovers and surveyed this peaceful scene. The other driver and the crews, deep in their funks, stayed inside the vehicles. In front of us were two shallow circular pits, which looked as if they might have once been deeper but had slumped in. They certainly didn't look natural. We walked over to examine them. Thomas, following, said, "Those are da pits to catch the animals. Da Old Ones make dem, long time ago. Den dey herd da animals into the valley and push 'em dis way. Da animals, dey try to get away, dis way," he said, gesturing to the road we had come in on, which followed a narrow break in the rocky cliffs. "Den de animals fall in de pits."

"Oh. So these were animal traps made by Bushmen?"

"No, not by Bushmen. By da Old Ones. Before da Bushmen come."

Since the crews were too spooked to be of any use, Teddy and I unloaded the gear ourselves, and hauled it a hundred yards or so around the perimeter of the big kopje to the left of the road. It was a solid mass of a very hard mafic gneiss. In this climate, there was scarcely any weathering: The rock rose straight from the sand at a steep angle, with just a few rounded boulders forming a rubble pile at its base. We tested the rock here and there with our hammers. It was, everywhere, of that pure, bell-like quality that we had found at Toteng Star. This was going to be a site of equally superb sensitivity.

We installed the geophone and prepared the site carefully. The gain was again an impressive 120 dB. After setting up the two recording drums for eight days of unattended operation, we carefully battened everything down, burying the cables with large rocks and building strong shelters for the geophones, batteries, and recorder as extra precautions against baboons that might live in the rocks. Because of the signs we had seen that there might be some humans around, we swept a clear area in the sand around the instrument and left signs that warned against tampering, in Setswana and English, on a perimeter of rocks we made around the swept area. This is usually the best way of keeping people out, particularly "primitive" people who are usually afraid of strange and unusual objects. When we were very satisfied with our installation, we turned and walked back to the Land Rovers.

"Oh, my God!" I received a shock like a chemical rush. It was actually accompanied by the sensation of the hair standing up on the nape of my neck. Teddy and I stood stock still, staring at the sight before us.

Thomas was standing in the clearing in front of the Land Rovers, talking animatedly, in the clicking tongue of the Bushman, to what looked like a Bushman chief, behind whom, on a low rise, were squatting some forty Bushmen in two rows, clad only in loincloths and holding spears upright, butt ends on the ground. They were

staring straight ahead at Thomas and the Land Rovers with fixed
and unblinking expressions, muscles bunched and ready for action.

I walked up to Thomas, but before I could utter a word, he said,
out of the corner of his mouth, "Get in the Land Rover." I did what
he said. Teddy and I sat in our Land Rovers, watching Thomas continuing the negotiations, not understanding a word of what was
being said. In back, the crews were lying, with their arms covering
their heads, whimpering pitifully. The Bushmen, in their ranks,
didn't move a muscle. Finally Thomas returned, climbed in, and we
started up. We turned around and drove back down the road, and a
sense of relief rose off us like a blanket. Not a word was spoken all
the way back to camp, each of us being left to our private thoughts.

That evening after dinner, Teddy and I opened up a fresh bottle of
scotch and had a few drinks. Suddenly he started to giggle.

"He, he, he, he. A Navaho with a rifle. That I can deal with,
Scholz. But Bushmen with spears? He, he, he."

We settled back into our usual routine in the week that followed.
Each day we made our rounds of visiting the instruments—changing batteries and the paper records on the drums. There was a lot of
technical upkeep to be done, so Teddy was kept busy working in
camp on the instruments while I took care of more mundane matters in the field. The problems with the instruments were mainly
caused by the heat. A lot of the more sensitive components were
simply being heated above their normal operating ranges and mal-
functioning. Teddy was a wonder of resourcefulness in that kind of
work. I had installed one station near an abandoned cattle post, a
wattle hut within a kraal made by piling up thorn tree branches to
form a spiky enclosure. It was a dismal location, without a bit of
shade anywhere around. I had rigged up a tarp shelter to protect
the instrument from the sun, but this had turned out to be a bad
idea. A few days later I returned to find the tarp had blown off, and
burned my hands trying to open the instrument case. It was hot
enough to fry an egg on. The heat had damaged some component in
the digital clock; it was fibrillating in wild oscillations. I brought it
in to camp and showed it to Teddy. The clock was a solid-state de-
vice, hermetically sealed in a soldered stainless steel can. It was not
the sort of part that was meant to be repaired, but since we had no
spares he went to work on it anyway. He cut through the can and,

excavating into the potting compound that filled the interior, un-earthed the burnt-out capacitor that had caused the failure. He re-placed it with a near equivalent that he found in one of the many boxes of odds and ends he kept in his repairs kit, and the clock was up and running again. This was quite a feat: the electronic equiva-lent of battlefield open-heart surgery.

On another occasion he returned to camp later than usual and, pulling up near the Bedford, beckoned Blackie over. "Blackie, we blew the diff on this Land Rover. What are the chances of finding another one in Maun?"

"What? What happen, Mr. Ted?"

Ted pointed under the rear of the Land Rover, where we could see a hole in the differential casing, still dripping oil. "Differential. Broken. Must be replaced. Understand?"

Blackie's eyes widened with amazement. "If this break, how you get back to camp, Mr. Ted?"

"Disconnected the drive shaft and came home on the front axle." He opened the rear door and showed us the drive shaft lying in the back of the Land Rover.

Blackie's eyes suddenly registered comprehension and a wide smile spread across his face. "Very good, Mr. Ted. Very good. Most good." He chuckled, wagging is head in delight. Blackie was im-pressed. I had never seen that happen before.

As we drove around Ngamiland doing the rounds, I had plenty of opportunity to question Thomas about the bushmen of the Kgwebe Hills. It seems that they were not "tame" Bushmen, of the type I had seen working as serf labor on the cattle posts on the fringes of the Okavango near Maun. Neither were they the true "wild" Bushmen that existed as hunter-gatherers in the central Kalahari and were seldom seen by outsiders. This was a tribe that had taken up the agrarian life but had managed to retain their independence, settling in the Kgwebe Hills and maintaining there a sort of tribal fiefdom. Thomas was not very clear about when this had happened—he often had a vague sense of time—but he said that no outsiders had visited that place for years, not even agents of the Botswana gov-ernment, which preferred simply to leave them alone. I could learn no more about them than that. But on the question that most concerned me, he was emphatic: They would not touch our instrument. How he had reached that agreement with them he would not say, but he did say that they knew what earthquakes were, and that our instrument was listening to them.

Underlying all of our thoughts that week was the knowledge that at the end of eight days we would have to return to the Kgwebe Hills to retrieve our instrument. As that day approached, the tension perceptibly built up among us, though no one broached the subject directly. In the meantime, though, the instruments at Toteng Star and in the U.S. Steel trench had co-recorded more than a dozen good solid microearthquakes. All we needed were the data from Kgwebe to make this week a great success.

We left camp on the appointed day with exactly the same complement as before, Teddy and I each in a Land Rover with a driver and two laborers. This time we drove silently, stoically, and a little grimly up into the Kgwebe Hills, past the burnt-out Boer farmhouse, and into the little valley. The whole place seemed unnaturally silent. Not a soul was to be seen.

Teddy and I approached the instrument with a mixture of eagerness and trepidation. With relief, we saw that nothing had been touched. I opened the recorder case.

"Oh, shit!"

The recording needle was vibrating madly on the drum, clipping from stop to stop as though responding to some violent agitation like a magnitude 8 earthquake. The needle, whirring across the drum, had scraped off the lampblack that we had carefully deposited on the paper as the recording medium. I shut it off. The whole record, for eight days, had been obliterated by these wild needle oscillations. There was not an intelligible fragment left on the whole record. But the history was clearly written there: The recorder had gone berzerk within an hour of our having left it there, eight days before.

"What the hell is this, Teddy?" I said. This was a technical problem, in his bailiwick.

We checked it out. The geophone, the amplifier settings, the panel meters, the circuit board connectors, everything looked normal. We carefully combed the area. There was no evidence that anything had been disturbed. There were no tracks within the area we had swept, no marks on the case, and the cable, the geophone, and the batteries were all in place and properly connected. Everything looked okay, except that something was obviously seriously amiss.

"What the f___? Over," Ted said, with the intonation of a radioman. This was his extreme statement of technical perplexity.

All this hassle, and then the instrument malfunctions. All that data lost, and there was never going to be another chance to revisit this site and do it again. I was stunned, drained, and appalled, all at the same time, and Teddy knew it.

The crew was happy to help us this time as we dismantled the gear and packed it back into the Land Rovers. As soon as we cleared the hills, they were back in their usual garrulous spirits.

Immediately after we got back to camp we took the gear into the instrument tent where Teddy went to work on it. I was still so steamed up I went for a walk in the bush to cool off. When I got back, still in a glowering mood, I opened a beer and sat down at our mess table and stewed. A half hour later Teddy emerged from the instrument tent and gave me a peculiar look.

"You're not going to believe this, Scholz."

"I'd better believe it. I sure as hell want to know what went wrong out there."

"Well, ah, what can I tell you? There's nothing wrong with the unit. I set it up and it's working perfectly now."

"Ehh? Come off it, Ted. The thing went bananas for a week. It was in bloody open-loop oscillation. You saw it. Don't tell me there was nothing wrong with it."

"Cool off, man. Yeah, I know, I saw it. But I've just been over the unit from one end to the other, and everything checks out. I'm telling you, there's nothing wrong with it. Come on into the tent. See for yourself."

We went into the tent. The recorder sat on the bench, tracing a nice clean straight line on the drum. I stamped on the ground, and a wiggle appeared on the record.

"Well, so what gives, then? The Bushmen must have messed it up."

"No way. I mean, even if they had messed around with it and covered up their tracks later, there was no way they could have made those oscillations. At most, they could have disconnected something, which would have just left a dead signal. Whatever made those oscillations, it wasn't in the electronics. And whatever it was, it was still going on when we were there, when everything was hooked up, because we checked."

"Well, then, what's your explanation?"

"Damned if I know, Scholz. I've never seen anything like it."

"Then what am I supposed to think, for Christ's sake? That Bushman spirits zapped our data?"

"Hey, don't ask me, man, I only deal with electronics. But I can tell you one thing that's funny about this whole deal. When Thomas saw that record, he didn't look at all surprised. He looked like he was expecting it, or something."

8

It was while we were working out of Toteng, remote from any town, that our party solidified into a tight working team. Our daily routines were very simple and indistinguishable from one day to the other, because we took no days off except once in a fortnight or so when we went into Maun for supplies and a bit of R & R. Every morning Deacon would wake us at 6:30 with coffee, which he now made the way I had taught him—stirring the grounds into boiling water in a coffee-can billy, then, after a few rolls, removing it from the heat for a minute to steep, finally settling the grounds with a sharp rap of a spoon on the side of the billy. We drank our coffee in the tent, while listening to the BBC World Service on the short-wave. There was a program several times a week at that hour called *Letter Box, with Margaret Howarth*, in which letters from listeners would be dramatically read out by actors to said Margaret, who would then reply to their various concerns and complaints. A typical letter would be from somebody living on some plantation deep in the wilds of Borneo writing in complaining about why the radio program *What-ho, Jeeves* had been shifted to three in the morning his local time, and how upsetting it was to have to get up every night at that hour to listen to it. The actors would read these out so earnestly that we would get a howl out of it, and make up our own imaginary letters to Margaret (to which the other would do her replies).

After we had a washup, Deacon served breakfast, which we ate while listening to the BBC World News. The food problem had by this time been more or less solved. Deacon, who we had found out was called that because he actually was a deacon in his church, turned out to be not such a bad cook after all. He simply cooked whatever his clientele wanted. If they wanted badly cooked British food, that's what they got. He quickly learned how we liked our food, and we taught him our various favorite recipes that could be made with the limited variety of food stocks available. This included Teddy's famous "secret" teriyaki steak recipe, which turned out to be lumps of steak marinaded in soy sauce and garlic and then sautéed (actually, this was not his "true" secret recipe, which included lemon, but there was none of that to be had out there). Deacon was actually quite a versatile cook, considering what he had to work with. His cooking utensils consisted of a frying pan, a set of three-legged cast-iron kettles in several sizes, some sharp knives, and a couple of big forks and spoons. The kettles were in the classic boil-up-the-missionary style, though he had none quite so large. Deacon used these for everything from cooking stews to baking scones. All the cooking was done on a fire that consisted of three or four long poles laid with their ends together on the ground. He kept this fire going all day long, regulating the heat by kicking the pole ends closer together or farther apart. I never saw him light it.

The BBC News ended at 7:30, which was the time assigned to the Botswana Geological Survey for radio-telephone communications, which we were obliged to tune in on every day. This usually ended up being a gossip session among the clan of Afrikaner well drillers, but occasionally a message for us would be relayed from Lobatse, which we would have to pick up on and acknowledge. We were also obliged to notify Lobatse whenever we changed camps or sent vehicles in for repair or exchange.

At eight, Thomas and Samuel would pull up by our camp with their helpers in the two Land Rovers. We would issue brief instructions, the gear needed for that day would be loaded up, and we would be off on our daily rounds. Except for the particularly remote sites, we would normally visit each instrument daily, to change the smoked paper on the recording drums and generally to check that everything was going OK. There would be considerable anticipation when opening the recorder box to see if anything had been recorded overnight, but generally after the first one of the day you pretty well knew the answer: not much. But there were those exciting days when an event was recorded, and you hustled to the next instrument to see if it had also registered it clearly.

One chore that was unusually vexing was putting radio time marks on the drum, which had to be done when first installing the new record and removing the old one each day. This was done to calibrate the clocks. The clocks we were using were the first generation of solid-state quartz clocks, and they still drifted quite a bit, especially under the extreme thermal variations that they were undergoing here, from 120° at midday to 60° or so in the early morning hours. In those days before the SATNAV systems, the usual technique was to pick up the digitally coded time pips broadcast on shortwave by the American WWV system and play these into a special amplifier which would record them on the drum. The problem was that in the Kalahari, shortwave reception was almost nonexistent for six or eight hours in the middle of the day. What you would generally end up doing was to run out a couple of hundred feet of antenna wire, and have the crew walk this around though all points of the compass while you searched through all five of the shortwave bands on which WWV broadcasts, hoping to pick up a signal good enough to record. This was a time-consuming procedure with no guarantee of success, since you were basically relying on some fluky ionospheric fluctuation to give you a few minutes of good reception. The alternative was to tune in Radio South Africa on the AM band. They played time pips for the first five seconds of each hour. We could usually pick that up with only a fair bit of static. The problem was that if there was static or some other interference during that crucial five seconds you would have to wait another hour, all the while listening to some of the dullest programming in the annals of broadcasting.

If we got back to camp early in the day, there was an alternative to RSA for radio entertainment. This was provided by Radio Lourenço Marques, a pirate top-forty station in Mozambique that beamed into South Africa and leaked everywhere else. The omnipresent dee-jay on RLM called himself Wolfman Martin and did a lame imitation of Wolfman Jack of Tijuana radio fame. What would you expect of someone who calls himself Wolfman Martin? His howling was wimpy. Like all top-forty stations, this one repeated itself in a one-hour loop over and over again, with a script that varied about as rapidly as a television soap opera. Whenever Teddy was working in camp, he would play RLM all day long, which if I was around would drive me batty. I can date my time in the Kalahari with great precision. It was when Eric Clapton made a hit with his rip-off of the reggae tune *I Shot the Sheriff.* I must have heard that hateful song a couple of thousand times. It was such a relief when the short wave came up in the evening and we could get the BBC again.

Our diet was based on beef, mealies, and whatever tinned and packaged goods we had bought on our outing to Mafeking. We soon found ways to expand on this limited fare. Guinea fowl were plentiful in the scrub forests surrounding the lake shorelines, and Thomas agreed to shoot these for the pot if we would supply the ammo. On the next trip to Maun we bought a couple of boxes of shotgun shells and let him have a go at it. Thereafter, whenever we spotted guinea fowl, he would stop the truck and go after them. While hunting he would present a ludicrous figure, pursuing his prey through the brush with a hippity-hoppity tip-toe, with his bony knees pumping up and down between his enormous boots and shorts and with his antique shotgun tucked under his arm. But no matter how bizarre he looked to us, he seemed invisible to those normally skittish birds, and his stalks were usually successful. Not wanting to waste a shell worth 22¢ on a single bird, he would hold off until he could get several lined up. The most I ever saw him bag with a single shot was five. He considered it a failure if he got only a single bird with a shot. Samuel had a shotgun as well, and wanted his chance to try. But he couldn't hit the broad side of a barn, and after awhile we stopped giving him ammo.

The only inconsistency in Thomas's hunting was his gun, which was constantly misfiring. Many a time we would watch him sneak around a bush and draw a bead, and then hear a disappointing "click." Finally Teddy asked to see his gun, which he took away to the instrument tent. It was a classic old hunting piece, a single-barreled hammer-action Remington stamped "1906." The stock had broken and been mended with copper wire, then wrapped around with python skin, much in the way we might do with fiberglass tape. Teddy extracted the firing pin and showed it to me. It was worn to a nubbin. He rooted around in his various boxes until he found a piece of steel of suitable size, and with a hacksaw and file proceeded to make a new firing pin. He reassembled the gun and duly presented it to Thomas, who tested it out. Bam, bam, bam, three shots and no misfires. Teddy went up another few notches in the esteem of the crew.

Besides supplying the ammo, we contributed to the pot in another way, by bream fishing. Our favorite spot was a ford in the river several miles upstream from the bridge. There were some nice riffles and pools there that made for ideal bream habitat. With our light spinning rigs we could catch a half a dozen or so fish, each a bit over a pound, in half an hour. This was quite a treat for the crew, who had never eaten fish before. Thomas had to show them what to do with them. He even taught us how to make a fish batter with mealies, which is a kind of cornmeal. Teddy made up a wire rack and taught Deacon how to smoke them, so we could have kippers for breakfast. So with the addition of both the fish and the guinea fowl, our diets became much more interesting and varied. The guinea fowl is the same bird that the French grow domestically and market as *pintade* , but the wild bird is much more flavorful. It tastes like gamey chicken, and is good prepared any way you would chicken.

The layout of the camp at Toteng was exactly as it had been in Maun. This layout in some sense symbolized our relationships. The crew worked with us but did not live with us. They lived in a different neighborhood, so to speak. This was a sensible arrangement, considering the enormous cultural and economic gulfs between us. The R10.50 per diem subsistence allowance the U.N.D.P. was providing us was a week's pay for one of the laborers. Their allowance was something like 50¢ a day. Deacon, by dint of his duties, had a crossover role to play, associating with both us and the crew in camp. Hence his work station was situated between the two camps. Deacon was a jolly character and got along with everyone, so he was well-suited in this role, although he was often the butt of good-natured jokes by the crew. The position of Thomas within the crew's hierarchy was anomalous, since he was not a member of the Botswana Geological Survey like the rest of the crew. However, his role as my personal driver and guide provided him with considerable status in the crew, and he also earned a lot of respect just by virtue of his personality and knowledge. Blackie was the paramount chief of the crew. He was a man of great dignity and gravity in bearing. His stature was as great as his authority; he towered over the others. He was bosun to my skipper, though the metaphor is not quite right; his authority was greater than that of any bosun.

The division of labor within the crew was quite well defined. If we had a puncture, which was often, because the thorns from the thorn trees were continually working their way through the tires, Thomas and I would retire beneath the nearest tree while the laborers did the repairs. The driver drove, a skilled profession. He did not engage in common labor. A similar division existed among the laborers. The senior man did the repairs, his junior the fetching and carrying. All tasks were preassigned and were carried out without dispute or discussion.

We seldom had any social contact with the laborers except at the midday break. These were somewhat limited because of the extreme difference in our backgrounds. I recall one such exchange with Robert, when he asked me where we came from.

"From New York. It's a big city in America."

"How long does it take to come from there?"

"Oh, it takes about a day and a half by plane."

"But how long by Land Rover?"

"You can't come by Land Rover. You have to cross the ocean. It is a very wide place full of water."

"Oh. Is it as wide as the Zambezi?"

The exception was a young fellow named Straing, who was in Teddy's crew. He was very curious about everything we were doing, and Teddy, a born teacher, was encouraged to try to make a technician out of him. While the other laborers' tasks with the instruments were limited to digging, chopping, and fetching and carrying, Teddy was teaching Straing how to play in the radio signals and set up the amplifiers, and during lunch breaks he was teaching his eager pupil the rudiments of how the instruments worked. In the evening Teddy would proudly report to me on the progress of his charge. This whole business had an unhappy ending.

One evening Straing appeared in our camp, tears streaming down his face. We were eventually able to work out through his blubbering that Blackie had fired him from the crew, and he was being sent back to Lobatse the next day. "It because my working with Mr. Ted," he stammered.

He begged me to talk to Blackie for him. I knew this to be a breach of camp protocol, but I felt I had no choice but to have a word with Blackie. When I entered the crew's camp they were all sitting around the fire having a cup of tea. They looked startled and uncomfortable at my presence. Blackie glowered, and said, "What you doing here, boss? You shouldn't be in this camp."

"I'd just like to speak to you about Straing, Blackie."

"Straing's behavior not proper for young man. This bother crew very much. He must leave. So tomorrow he goes back to Lobatse."

That was the final word. There was no doubt about Blackie's authority to make such a decision, and I could see that there was a kind of logic behind it, too. Still, I was sorry about the whole incident. I was sorry for Straing, and sorry for the country. Straing was just the sort of person who could be trained in the type of technical skills that were in so short supply in this country, but he wasn't permitted to do so simply because it wasn't perceived to be his place.

The Ngami basin was used sparely by a few scattered cattle posts, but there was still a fair bit of wildlife to be seen. There were amazing flocks of birds, starlings, I took them for, that in flights of a hundred or more would bank and turn and dive in tight formation, suddenly in a flash changing the appearance of the flock like the flitting of a school of tropical fish. If you came upon them roosting, they would skirr away in a thrumming rush, startling in its abruptness. Ostriches were common, too. Sometimes we would come upon groups of them, all standing in a circle with their heads together like café chairs tipped against a table in the rain, holding some kind of secret ostrich palaver. There were a few small antelope, duiker and steenbok. The first time Thomas pointed out a steenbok to me, I couldn't make it out at first. It was a tiny little antelope, like a toy, only a foot and a half high at the shoulder with delicate spike antlers. It stood staring at us from next to a tree stump, then darted away. Thomas wanted me to buy him some ball shot so he could shoot one for the pot, but I never did it.

The nights were quiet, the silence broken only by the occasional distant wail of a jackal. I discovered the splendor of the Kalahari night on our first evening at Toteng when I walked out of camp to do my nightly business. On the edge of the bluff I looked out over the desert. I could see the entire Ngami basin spread out before me for many miles. I thought, "What a beautiful moonlit night," and then, with a start, noticed that the moon was not up. It was a starlit night, and so bright that I would have no trouble making my way back to camp through the scrub woodland. I saw the southern sky that night as if for the first time. The Magellanic Clouds, the two galaxies closest to the Milky Way, glowed overhead with so many stars they could not be distinguished except for a few bright sparks within their pearly glow. I recognized the Southern Cross down on the horizon. The southern sky is so much more spectacular than the northern, and there is probably no better viewing place than the Kalahari. After that I went out to look at it almost every evening, and never lost a sense of awe at the sight.

And then, on the other hand, there were the snakes. We first encountered them when we were working in a very brushy thicket. Teddy was bent over beneath a bushy tree, doing something with an instrument, when suddenly Robert gave a big swipe with a stick right over his head. "What the ...?" I said and then noticed a snake flying through the air. Ted went and picked it up behind the head with his thumb and forefinger. He held it up, wiggling around his arm. It was thin and a nondescript brown, no more that two feet long. "Cute little bugger, huh?" said Ted. Thomas said, "Dat twig snake, Mr. Ted. Very bad snake. He kill you most quick." Teddy turned white in the face and tossed the snake into the bushes, where Robert pounced on it and beat it to death with the stick.

When we left the thicket a few minutes later, we encountered a cobra on the path. The sight of a cobra suddenly rearing its hooded head up a few feet in front of you is enough to give anyone a good long-lasting case of nerves. I was used to working in rattlesnake country out west, but this was an entirely different experience altogether. Rattlesnakes are pretty considerate as far as snakes go; if they have a chance, they will warn you by rattling when you get too close. Besides, they live on the ground, so it's easy to get used to scanning the ground in front of you when you are walking in rattlesnake country. But here the snakes live in the bushes and the trees, too. Thoroughly spooked, we found ourselves walking at a snail's pace through the bush, nervously examining every tree, bush, and patch of grass for a hint of any of the twenty or so species of snake that could be found there. We seemed to have been treated to a plague of snakes on that day. In addition to the twig snake and the cobra, we also saw a puff adder and a black mamba. It was as if as soon as we had become aware of snakes that we started seeing them everywhere.

The moment we got back to camp I shuffled through my book box and pulled out a copy of *Snakes of Africa* that I had picked up somewhere along the way. From my somewhat obsessed reading I learned several salient facts about African snakes. The first was that all snakes in Africa are extremely bad news. Their venom comes in two general varieties: those that attack the central nervous system, like the mamba's, and those that decoagulate the blood, like the boomslang's. Some have a mixture of both. They are all equally toxic and the symptoms they produce do not bear repeating. It is a question of where you get bitten that determines how long you will live without treatment; it could be from a half hour up to several days. The second important fact is that the only treatment is by injection of a specific antitoxin, and that doesn't always work.

I went over and got one of the snake-bite kits that we had so casually thrown in the lockers of the Land Rovers the month before. It was packed with syringes and vials full of evil-looking brown liquids. In front was a thick instruction booklet, which I flipped open. The first page had a flow chart for identifying the snake that had bitten you. You had to identify the snake to know which antitoxin to inject. On the second page I was brought up sharply by the following announcement:

"These equine antitoxin sera should not be administered to those with a history of asthma or other chronic allergy, since asphyxia from allergic shock may result in death."

"Well," I thought, as I lifted my inhaler to my lips, "I guess I'd better give the snakes a wide berth."

The trip into Maun from Toteng took about two hours. We usually went in on Saturday morning, early enough so that we could order meat at the butchery, a big white shed at the edge of town. Our order was big enough to induce them to procure a steer which they would herd up and slaughter around noontime, so we were assured of getting fresh meat in the afternoon. They offered three cuts: filet, sirloin, and the rest of the beef cut up indiscriminately into cubes and sold as "ration meat." Ted and I would buy a filet and the crew stocked up on the ration meat. We would then stop by the little farm tended by the inmates of the jail to buy whatever produce they had available, sometimes pumpkins, at other times onions or beans. For other goods we would visit one of the general merchandise emporia. The one run by an Indian family was my favorite. It was one big open warehouse with an impossible jumble of every variety of goods stacked in piles at the back. You would put in your order at the front counter and a clerk would go back and search through the piles. Inevitably they would come up with something that was a reasonable facsimile of what you were looking for, but it was a wonder how they were able to find anything in all that disorder. Behind the counter was an big old wooden apothocary's rack with pigeonholes filled with an enormous variety of small bottles of antique remedies such as Lydia Pinckham's Tonic and Dr. J.S.V. Brown's Genuine Indian Root Oil Medicine. Most of these concoctions, which tended to be primarily composed of alcohol tinctured with a few other drugs like cocaine and wormwood, had been illegal for sale in the U.S. since the first Food and Drug laws were passed in the '20s, but seemed to be doing a brisk trade here. Old men would come up and buy one of these miniature bottles and quaff it down at the counter, and then order another.

It was usually Riley's for lunch, but from time to time Dave Potten would invite us to lunch at his bungalow. Lunch with Dave and his wife, Sarah, was always a delight. They had both come out to southern Africa from England five years earlier and had fallen in love with the place. Sarah was a nurse in the local hospital, and with Dave's job at the U.N.D.P., a prized position, they were well set up in Maun, but their ambition was to buy a farm in Zimbabwe once the war there was over, and to raise a family there. This was a shocking idea to me, because no matter how beautiful that country is, the long-term future there for the white minority, once the black government was in power, seemed to me to be an open question. Still, you couldn't resist their cheerful optimism. They were the first Europeans we had met in Botswana with this attitude, and it was refreshing. I've often wondered how it worked out for them.

The thing about Riley's on a Saturday night, or any night for that matter, was that you wanted to be elbows up at the bar at five o'clock sharp when Ronnie, the head bartender, opened up. This was because Riley's had a single refrigerator, of the old type with the condenser on top, which was stocked with a couple of cases of beer and two trays of ice cubes. That gave you about a five-minute leeway to get a gin and tonic with ice in it, providing they had any Schweppes, of course. After the first twenty minutes, the cold beer was gone, too, so the rest of the evening you drank the regular warm beer from the storage room. The two popular brands were Castle, a South African beer, and Carling's Black Label, the old defunct Milwaukee brew, which I suppose had found this market niche on the strength of its name.

The bar at Riley's is a long room with the bar on one side, a set of doors leading to a verandah on the other, and a few tables scattered here and there in between. There is sometimes enough space to sit at the tables, though not on a Saturday night. The place to sit was on the verandah, which overlooks a scruffy garden separated by a low wall from the parking lot and town square on one side and a marshy inlet of the river on the other. From the verandah, in the early evening, you can watch the sun go down and greet new arrivals, because everybody at Riley's knows everybody else, whether or not they know their names.

Teddy, however, had taken to be a bit furtive at Riley's, and it was no longer possible for us to sit on the verandah. This was because of his troubles with the Groovy Queens. They were a group of teenage girls who hung out on the verandah, particularly on Saturday nights. They were Maun's version of mall-ettes, the sort of teenage girls who hang out at suburban shopping malls in the States. Their main activity was pestering the young men going into the bar, and particularly trying to cage them into buying them Groovys, a South African brand of soft drink. Since they weren't allowed into the bar itself, they would station themselves in groups on the verandah, where anyone who attracted their attention would have to run their boisterous gauntlet in going in or out, such traffic being assured, since the beer was inside and the toilets were outside around back.

Teddy's problems with the Groovy Queens were to a large extent his own doing. It all started on one of the first evenings we spent at the place. Sometime in midevening, in that period of equilibrium long after the bustling for first beers was forgotten and before the anticipation of last calls stirred, there was a tremendous commotion of feminine squeals and screeches from the verandah. We all rushed out to see what was happening. There was Teddy, in the dark, demonstrating his electric yo-yo to a gaggle of screaming Groovy Queens. And what an exhibition it was, too. He must have been secretly practicing. He did sleepers, walk-the-dogs, rock-the-baby, and finished with a spectacular triple round-the-world, all glazing the retinas with stroboscopic arcs of flashing red, green, and white light. What pyrotechnics! I had never myself seen quite such a bravura yo-yo performance before, and the Groovy Queens went absolutely ga-ga over it.

Teddy didn't stop there, though. One late afternoon shortly afterward, we were having a beer on the verandah, enjoying the setting sun, when Teddy called one of the Groovy Queens over to our table. He took her by the hand and said, "Let's see what we have here," examining the jewelry that ornamented her wrist. It was the usual stuff, a couple of bracelets composed of strands of thick copper wire, topped by a stack of black rubber O-rings. O-rings, being the modern substitute for giraffe-tail bracelets, which they resembled, were the most popular ornament worn by the girls. The genuine articles were no longer available at prices affordable to ordinary people. This was the cause of our perennially leaky jerry cans, from which the crew were forever filching the O-ring seals to gain favor with the local girls. "Let me show you something," said Teddy, and gathering up the beer can pop-tops from our table, proceeded to make her a bracelet of them. She went cooing over to her pals to show it off, who oohed and aahed and came rushing and gushing over to get theirs. Teddy went into production. Scrounging a pile of discarded pop-tops from the bar, he sat down and on the spot proceeded to create the Teddy Line of pop-top accessories. It included bracelets, with and without bangles, belts, elaborately woven necklaces, and a tiara to round out the collection. This created the fashion sensation of the season for the young ladies of Maun. Pop-tops were a readily available commodity, most of the metal in Maun being in the form of fuel drums, bits of old Land Rover, or beer cans. Within a week all the young women of the town were decked out in pop-top art.

So this was how Teddy became a celebrity among the Groovy Queens. Soon he had his own fan club, a group of eight or ten girls who clustered around him, giggling, whenever he appeared at Riley's. Teddy, ever the flirt, reveled in all this female attention, but it wasn't long before things began to get out of hand. Inevitably there arose a competition within his fan club as to who was going to be his top fan, his number one girl, the one who had the primary claim on his affections. After considerable bickering and not a few hairpulling fights, one girl, tougher and more determined than the others, emerged as Teddy's uncontested Number One Girl.

It soon transpired that Number One Girl's ambitions did not stop with official fan club activities. One afternoon, when we were still in our camp at Maun, we arrived after our swim to find Number One Girl and one of her pals ensconced in our tent. Number One Girl greeted Teddy effusively and her pal gave me a coquettish fluttery smile. Uh-oh.

"Deacon! What's going on here, Deacon?"

Deacon emerged from behind a tree, where he had been tactfully lurking. "These girls want to marry Mr. Ted and you, boss. Very nice girls, boss. They be good wives."

"Deacon, Mr. Ted and I don't want to marry them. We are already married. We have wives, Deacon, back in America."

"No, no, boss. These be camp wives. Take very good care of you. Keep camp very clean. After you leave, they go back to village. No problem, boss, dey be very good girls." Deacon was enthusiastically warming up to his matchmaker role.

"No, Deacon. No, no, no. We don't want wives. No girls in camp, Deacon. That's my rule, no girls in camp. Understand? Please tell them to leave. Now, understand?"

The girls were staring at Deacon and me, transfixed, wanting to know what we are saying, since they didn't understand more than a few words of Pidgin English. Teddy was looking very uncomfortable. Deacon was looking at me as if I had some previously unrecognized mental disorder. "OK, boss," he sighed, "I tell them go home." He went over and got into a long harangue in Setswana with the girls, to which they responded with much beseeching and gesticulating. Teddy and I grabbed a beer out of the fridge and went and sat under a tree. It was twenty minutes before they finally left.

Number One Girl was not so easily brushed off, however. She appeared several more times outside camp, where she beckoned Deacon from the bushes with new arguments for him to transmit as to why she should be accepted into our tenthold. On the last of these occasions, Teddy, who was looking a bit haunted by this point and had clearly had too much, went and screamed at her to go away.

I must admit, however, to not being entirely innocent in the ultimate crisis with the Groovy Queens. This happened on our last night at Riley's before we left for Toteng. Joe had been going on more than a few times about having been the 1956 Singles Darts Champion of Rhodesia, so I had challenged him to a few rounds. I figured that a lot of water had gone under the bridge and a lot more flab over the buckle since then, and his lip needed a little competition on the board. The Groovy Queens were in full chorus out on the verandah and the game was being continually interrupted by the girls shouting in at Teddy whenever the bar door opened, and by the waiter, whom they kept sending in to ask us to buy them Groovys. The waiter was getting frazzled by all this, since the Groovy Queens were bother enough for him without the extra disturbance caused by Teddy's presence. I was also getting extremely annoyed, so much so that I couldn't concentrate on the game, so I finally slipped the waiter a five-rand note and told him to buy Groovys for the lot of them, thinking that would buy us some peace and quiet for awhile.

A great uproar ensued when the waiter brought the trayful of Groovys out to the verandah. Unbeknownst to me, I had just done the equivalent of buying the company of ten girls for the night. I was not to suffer the fallout from this brash action, though, because it was not I whom the girls assumed was the provider of such largess. When Teddy, unsuspecting, went out to relieve himself a few minutes later, a riot broke out. We rushed out to see the cause of this disturbance and found Teddy being mobbed by the Groovy Queens, with Number One Girl valiantly trying to beat them all off.

We were able to rescue Ted that night, beating a tactical retreat to our room, with Number One Girl fighting a rearguard action. But every time since then, whenever we visited Riley's we had to stay in the barroom all evening and spirit him away to our room at closing time, like some rock star sneaking out the back door after a concert. One night Number One Girl even found out our room number and came tapping on the window late at night. Even in the barroom Teddy could not be at ease, since the regulars, who had been following these developments with hilarity for weeks, chaffed him unmercifully. So Teddy had become a bit twitchy, even sullen, so uncharacteristic of him, whenever the day came up for a trip into Maun. No one was happier and more relieved than he when time came for us to leave Maun far behind and move north to the Chobe.

9

We left Maun for the north very early in the morning. Barring mishaps, we had at least ten hours' driving to reach our campsite at Ngwezumba in the Chobe National Park. This was cutting it close, since we had to reach our destination early enough to get camp set up well before sundown, and there were no possible intermediate stopping places. For that reason, we had loaded up our camp in Toteng the previous afternoon and moved it up to Maun for provisioning in preparation for an early departure. Teddy and I had followed along later and had managed to slip into Riley's without tripping the Groovy Queen telegraph, and so had enjoyed a quiet and relaxing evening.

Crossing the bridge over the Thamalakame just after sunrise, our little caravan turned north, following the road that ran along the bank opposite the swamps. This route ran for several hours along the edge of the forest in a region of scrub grass and sand, much like the back-dune area behind a beach, but flatter. The road was often deeply pitted with the relics of mires left over from the previous rainy season, which we had to wind around on many detours, so the going was slow. At midmorning we reached the tse-tse fly control post, which marked the northern limit of the region in which eradication measures were being made to control this pest. It also marked the limit of regular human habitation, because it is impossible to raise cattle in tse-tse fly country. The trypanosomiasis they transmit with their bites is deadly to cattle. While a bane to humans, the fly, for the very same reason, was a boon to wildlife. North of this point the animals of the veld did not have to face the almost certain annihilation associated with cohabitation with men and their cattle posts. The tse-tse fly is the last line of defense for the wildlife of much of tropical Africa.

The control post consisted of an insecticide dusting enclosure, a large galvanized iron shed built over the road, and several buildings serving as living quarters for the workers. We had to stop there to register our vehicles. On the way out, they would have to be dusted with insecticide. While we were hanging around, Ted went over to have a look at the dusting shed. "Hey, Chris, come and have a look at this," he shouted. "You're never going to believe this." I went over and peeked with him through a crack in the corner of the shed where the galvanized sheeting didn't meet. Inside was a truck, surrounded by a half-dozen gauze-masked men in shorts, marching around the truck pumping away at small cylinders that emitted little white puffs of insecticide. Step-poof, step-poof, they went in unison, as though dancing a death dance to the fly. "Flit-guns," he tittered, "He-he-he. Flit-guns, would you believe it? I haven't seen one of those since I was a kid."

Opposite the dusting shed was a group of Bushmen who were offering various handmade artifacts for sale. I guess this must have seemed like a prime business location to them, but I couldn't imagine them getting more that two or three potential customers a day at this out-of-the-way spot. They had bows and arrows and various kinds of crude baskets on display. The bows were very simply constructed, looking like nothing more than a roughly carved tree branch. They were quite slender and small, well short of three feet in length. The first one I tried broke. The Bushman trader showed no reaction to this disaster. He simply offered me another one to try out. Using a gentler pull, I managed to shoot the little arrow fifty or sixty feet with this one, so I bought it as a souvenir, along with a few of the arrows. At the time I thought that it was just a home-grown version of shoddy tourist goods, since it obviously wasn't possible to kill an animal with such a weakly projected arrow. Later I was told that although mine wasn't a prime specimen, it was a genuine Bushman bow. It was not used to inflict a killing wound, as the European longbow. The arrow was used to deliver and inject a stunning poison.

We weren't more than a half-hour past the tse-tse fly control post when we started getting bitten by them. Teddy and I were riding with Samuel in the pickup, when Teddy yelped, "Ouch! Jesus!" and slapped his neck. I was immediately bitten, too. It was a very painful nip. At first we didn't know what was doing it. Then we saw the flies zipping around inside the cab. Teddy slapped one hard against the dashboard. When he raised his hand it just flew away, unfazed. We started slapping away at them, but usually missed, they were so fast. Samuel chuckled, "De fly, he very quick. When you slap at him, he already biting da next fella. He very tough, too." Trapping one in the corner of the windscreen, he held it up to us, and crushed its horny body between his fingernails. "Dis de only way to kill de fly."

We rolled up the windows and managed to kill all the flies in the cab, but after a half hour it was so stifling in the closed cab that we had to open the windows and endure the fly attacks again. When I had been bitten a couple of dozen times and had started to develop welts upon welts, I asked Samuel, "What are the chances of getting sleeping sickness from tse-tse fly bites?"

"Oh, don't worry 'bout the sickness," he said. "Dey say only 'bout one fly in ten thousand got de fever." At the rate I was getting bitten, that sounded like a dead certainty to me.

We soon passed the turnoff to the Khwai River Lodge, which was a posh air safari resort in the Moremi Wildlife Reserve, at the northern edge of the Okavango. According to the plan I had worked out back in New York, our last camp was going to be in that area. In the meantime we had a month's work to do in the Chobe, and my mind was completely occupied with that challenge, so I made only a mental note of the turnoff. The road deteriorated immediately as we passed the turnoff, since there was little traffic that continued on the way we were going. Leaving behind the more forested regions surrounding the swamps, we entered the vast dry veld at the heart of which was the Chobe National Park.

Just outside the park boundary we met a party of big-game hunters cruising for animals that might have unwittingly wandered across the unfenced boundary line. They were in what looked like an old Dodge Power Wagon, with a shooting platform mounted on top. It was a very macho-looking rig, like an armored car out of the Afrika Korps, which I'm sure was the desired effect. There were two clients up top, a man and a woman, who were trying to look very cool in their broad-brimmed safari hats, freshly pressed khakis, and long scoped rifles poking out at the ready. The hunter in the cab gave us a wave. The clients just gave us a long stare from behind their metallic coated aviator glasses, as if to say, "And whooo are you, barging in on ooour private safari."

At the park boundary there was a long pole across the road, from which a rusty slapped-up wire fence ran a few hundred yards off to either side, before gradually losing strands and posts until it petered out altogether. Next to the pole was a small hut, from which a couple of game scouts emerged after we roused them by pulling up in front and beeping. We presented them our various permission documents from the ministry. They looked these over carefully, turning them over and over in a pretense of reading them, then passed them back to us. With a snappy salute, they lifted the pole for us to pass through. These people had a lonely job, because no one is allowed to enter the park from the South Gate except for the occasional government-sanctioned party like ours, and although the provisions they made to prevent this seemed more symbolic than real, they probably saw very few visitors. The only public access to the Chobe is from the North Gate, at Kasane on the Zambezi, almost two hundred miles to the north. Those visitors are restricted to driving along a ten-mile stretch of closely patrolled road alongside the river. The vast interior of the park is completely off limits except for officially sanctioned parties. In the time that we were there, that consisted of our party and a group of game scouts camped in the Savuti region, about a hundred miles to the west of our camp.

The road continued north, skirting the edge of the Mababe Depression. This was a broad low-lying region which, according to Passarge's maps, had been well watered in the last century. It was now a region of dry seasonal grassland and pans. Its vegetation relied solely on the seasonal rains except in infrequent times of flooding. Then the Linyati swamps, just across the border along the Chobe River, would discharge into the Savuti Channel, and if the flow was strong enough this would eventually spill out into the Mababe, producing there a sudden luxuriant bloom of long-dormant flowers and grasses. The rains may be fickle in this part of the world, but in such flat terrain the drainage was doubly so.

We were traveling through a sandy region of mixed scattered forest and grassland. Off to our left we could catch occasional glimpses of the Magwikwe Sand Ridge. This remarkable feature, a fossilized dune, was sixty feet high, some two hundred yards wide across its flat top, and ran for more than thirty miles in a nearly straight line across the otherwise almost featureless plain. At its broader north end it abutted a parallel line of kopjes, the Gubatsa Hills. My guess was that the sand ridge had formed by airborne sand settling in the wind shadow of the hills, in much the way a tombolo, a linear sand spit, propagates out from the lee side of an island.

Our first encounter with wildlife was sudden and unexpected. Six sable antelope broke out of the bush to our left and at full tilt leapt across the road not thirty feet in front of us. It was a magnificent sight, with their bodies in midleap stretched full out and their long scimitar-shaped horns swept back along their gleaming black shoulders. Samuel slammed on the brakes and we narrowly missed the last of them, as another pair dodged and went across behind us. Breathless with excitement, Teddy and I belatedly scrambled for our cameras, as though it were important to obtain a photographic record of an image that was already, in all its dynamism, indelibly burnt into our memories.

Not much farther on, we watched a small herd of impala leaping in fluid motion through the brush, and a troupe of greater kudu which, holding their long spiral horns erect, stared haughtily at us from the camouflage of a thicket. Like tourists at a safari park in Kent, Teddy and I kept stopping and excitedly photographing the animals, as though they had been put out there especially to pose for us. This was an attitude that was to change very quickly when we encountered our first elephants. They were a couple of stray bulls, who uncooperatively kept ambling away through the bush as we attempted to photograph them. Teddy kept urging Samuel to get closer, so he could get a close-up shot with his Instamatic. Finally we got too close, and with a bellow of rage, they turned on us and charged. Samuel spun the truck around and floored it, and we sped off with the elephants thundering behind in hot pursuit. They gave up the chase after thirty or forty yards, but it had been a close call and we were badly shaken. The implications of where we were and what we were doing suddenly dawned on me. This was no safari park, this was wild animal country. There were something like ten thousand elephants in the Chobe, and countless other animals, and for the next month we were going to be living in their midst. Here the rules were set by the animals, not by us. We had better learn their rules, and quickly! I changed vehicles, climbing up with Thomas into my customary Land Rover. It was time to get some animal behavior lessons from somebody who knew something about the subject.

Coming out into the broad grasslands of the Mababe, we saw in the distance a vast mixed herd of antelope. Watching them through binoculars, I could make out roan and sable antelope, eland, the smaller gemsbok and springbok, and, in separate groups, zebra and wildebeest. In the foreground, under a lone acacia tree, lay a pride of lions. There were three or four females lying in the shade, with a clutter of cubs playing on their backs, while a big male dozed off to the side. Someone had lent me a 300mm lens for my Pentax for wildlife photography. It was one of the big old clunky ones, a foot and a half long and weighing five times more than the camera back, and I was just trying to learn how to use it. I fitted it, and steadying it against the Land Rover door, found that I could take individual portraits of the lions from forty yards. Bitten by photographer's greed, I got out of the Land Rover and walked toward them, looking through the lens and snapping pictures the whole time. The male raised his head and opened one eye, noticing me. As I approached closer, he repeated this several times. At twenty yards, he raised his head and let out a tremendous roar. "Okay. Fine." I turned tail and ran back to the Land Rover. The lion, seeing me retreat, didn't even bother to get up. I had just been given another lesson.

Continuing on, we came to a line of pans, similar to the one we had seen on our recon flight. They were circular depressions, several feet deep and ranging from fifty to several hundred feet across. They were largely dry, with just a little muddy water in the middle, and each was surrounded by a ring of trees. The whole area was heavily trampled and studded with the deep circular pits of elephant footprints. Here and there we saw a few elephants wallowing in the mud or browsing among the trees. These we carefully avoided. Thomas told me that the pans were made by animals, mainly by elephants digging and wallowing, and that they acted like shallow wells, collecting underground water. I could see that this was at least partly true, because they had slightly elevated rims, where the excavated sand had been pushed aside, and the water in them could not have been rainwater because it hadn't rained there in months. Following along the line of pans, we came to a group of them that were exceptionally broad and deep, and which were full of water. In them was living a group of about twenty hippos. I was astonished at this sight, because the nearest river was seventy or eighty miles away, across completely dry country. Thomas explained that the hippos would migrate out along the pans, and sometimes get stuck like this, where they would have to wait for the rains before attempting the trek back to the rivers. If that was true, then these hippos must have been in for a very long fast, because it didn't look as though there was much feed left for them around these isolated pans.

Teddy and I made our way, somewhat gingerly now, down to the water's edge to have a better look at the hippos. The water wasn't deep enough for them to be completely submerged, as is their usual custom, with only their ears and eyes protruding. The largest were only about two-thirds submerged, so they were constantly moving about to get their backs wet. One of these, finally getting fed up with our gaping, opened his huge tusky mouth and lunged at us, causing a tidal wave of a splash. We skittered back up the bank, while the crew laughed uproariously. Thomas, laughing along with the crew, said, "You got to watch dey's ears, boss. When dey wiggle dey's ears, dey mad. Den you got to watch out." Although none of the animals here was familiar with humans, and showed no particular fear of us, it seemed they all had a very well defined sense of propriety about their private space.

Early in the afternoon we came to the Savuti Channel, where it crossed into the Mababe Depression at the base of the Gubatsa Hills. There was still a good flow in the stream there, although five miles downstream it would be dry. Willows grew along the bank, their bright green foliage contrasting with the dry grayish growth we had been passing through for the last few hours. It was a pleasant spot, and we took our midday break there, splashing around in the stream to cool off before lunch. During lunch we met a party of the Savuti Game Scouts who had come down from their camp several miles upstream. They were surprised to see us. No one in the Chobe Park Service, they said, had been informed about our visit, even though it had been officially approved by the ministry long before. After we explained our business and told them where we would be camping, they were quite congenial and welcoming. They see very few visitors and were happy to get bits of news and gossip from the crew. Before they left, they asked us to look out for their water truck on our next leg transiting the park to Ngwezumba. It was bringing their monthly supply of drinking water from Kasane and was long overdue.

Our last leg was a long run to the east across the central part of the Chobe. Within an hour of leaving the Savuti we found the water truck, a small tanker, bogged down and high-centered in the sand. Its crew was struggling with a couple of shovels to dig it out. We all piled out and a half a dozen more shovelers got on to the job of digging out. Blackie and his helper, Boy, backed the Bedford up to the tanker and chained it up. The first couple of efforts of pulling wouldn't budge it, the weight of the water in its tank being so great. It took a good three-quarters of an hour of digging, winching with the pickups, and pulling with the Bedford, all masterfully directed by Blackie, before we prized it loose and sent it on its way.

With these various delays it was coming up hard on six o'clock by the time we reached our destination. We recognized Ngwezumba by the half-built brick and concrete bridge that was half tumbled into a muddy stream in the steep gully it was meant to span. After a couple of tries we found the proper line and got all the vehicles across safely, although we had a momentary worry that the top-heavy Bedford, swaying dangerously as it crossed the gully diagonally, might capsize. That would have been a disaster of first order. The far bank, where we pulled up, was a flat-topped and heavily wooded bluff. The road on that side was disused and partly over-grown, but it did afford a clearing for us to set up our camp. This was done with dispatch, with a great deal of hustling by all, because everything had to be set up, organized, and operating well before nightfall. When we finally settled in, Deacon fixed us up a hastily thrown-together but altogether satisfying meal. After dinner Ted and I had a whiskey and discussed the eventful day. We were feeling very satisfied with the success of the day's trek, and with our choice of a campsite. It was everything that I had hoped for. Well-treed, it would be shady and cool during the day, and the slight elevation of the bluff provided a panoramic view of the savanna off to the west, where we were now enjoying a magnificent sunset. It was so peaceful and my contentment was so full that I couldn't help but think that Hepworth was being an old biddy to carry on so direfully about this beautiful spot. Then night began to fall.

The first trumpeting came from the west. We strolled to the edge of the bluff to watch. The set sun still cast a fuzzy orange film over the rim of the plain. The sky above blended seamlessly from blue to black, with the sparkling jewel of Venus highlighting the transient but intense band of violet in between. An elephant herd, more than a hundred strong, was moving down from the north. We could clearly see their bulky but graceful forms as they left the forest and entered the open ground. They were trumpeting their greetings to another herd approaching from the south, which we could not see from our vantage point, but whose answering calls pinpointed them to us in the dusk. These images dominated the scene, but they were drawn within a broader canvas that was full of movement. In the half-light, it was almost as if the plain itself were vibrating, and those vibrations were themselves riding on longer and milder undulations. The plain was full of animals, moving in herds, all converging toward a single spot, the water hole, which I reckoned to be not more that seven or eight hundred yards to the south of us.

It started about a half-hour after dark. At first there was just a shuffling sound from the forest. This soon became more pronounced, and became recognizable as the sound of heavy scuffing footsteps in the sand mixed in with the rustlings of large animals moving through the trees. From one side there was a sudden sharp crack of a tree limb being broken, then another from the opposite side. We stared into the darkness but couldn't make out anything beyond the first line of bushes illuminated by our lanterns. Suddenly I realized what it was. A herd of elephants was coming down the old road on their nightly trip to the water hole, and finding our camp blocking their way, they were moving though the forest on either side of us.

The shadows we cast on the bushes suddenly increased in relief and, spinning around, I saw the crew in silhouette, frantically throwing wood onto their fire, which was now a roaring bonfire with flames four or five feet high. I yelled across to them, "Deacon, Blackie, how about some wood for our end?" Blackie's voice came back from out of the darkness, "Sorry, boss. Right away, sir." Three or four of the men came running over with big loads of wood. We quickly had our fire roaring, too. Our camp was on the forest side, so if any elephants were going to stumble into camp, they would do it from our end. They weren't as likely to do that as long as we had a good fire going to warn them away. And, of course, the fire, brightly lighting up the camp and casting its dark shadows into the forest, made us feel more secure too. There is something about a fire that makes you feel safe, even if you know it to be an illusion.

The elephant traffic picked up in intensity as the main part of the herd began to arrive. We now heard what sounded like whole trees being broken down, on both sides of the camp, as the elephants crashed through, breaking new trails around us. Every few minutes a shrill trumpeting would sound at close quarters, making us jump every time, but the elephants mainly communicated with rumbling liquidy muttering sounds, which were more unnerving because they were frequent and came from all directions, giving us some idea of the large number of huge animals stumbling around out there in the dark. The herd took more than an hour to go past. When the noises of their passage died down and we realized that our ordeal was over, we breathed a sigh of relief. This feeling was short-lived, because after a lull, a second herd began to arrive. Gradually, though, the movement of the elephants became more orderly and less noisy, with less trumpeting and tree breaking, as the new trails became more cleared out.

Meanwhile, activity at the water hole had been steadily picking up, creating its own din. What an amalgamation of noise was coming from there! It was a real cacophony. Dominating it all was the roaring of lions, which rumbled across the veld. This was accompanied by all manner of howls, wails, shrieks, yips, and barks of the other predators. I could recognize hyenas, jackals, wild dogs, and various kinds of cat. God knows what else made some of those sounds. They were all saying the same thing. Time to eat! The piercing and bloodcurdling death screams of animals being slaughtered testified to their success. Although the water hole was a good half-mile away, sound carries very well across the dry plain, and we were treated to the most intimate sounds of the feast, in all their gory glory.

There was a third catagory of nocturnal sounds which were disturbing in another way. These were low-pitched, almost subliminal sounds coming from around us in the nearby forest. The trouble with these sounds, which include the cough of leopards, and low rumbling *aaoouu* sound that lions use to communicate over short distances, the muttering of the elephants, and a few others that I never did identify, was that I could never learn how to tell how far away they were, nor could I often even be sure of which direction they were coming from. This creates a very disquieting sensation. These were the sounds that bothered me the most. It didn't take the most out-of-control imagination to think that the lion whose low moan you just heard is just on the other side of the tent canvas.

I live in New York City, so I'm used to hearing all kinds of racket at night. Traffic, sirens, motorcycle races, car alarms, boom boxes playing annoying kinds of music at full blast, people shouting in various languages, car accidents, jets coming into La Guardia, and in my neighborhood, occasional gunshots: These are all my normal sounds of the night. And like all New Yorkers, I pride myself on the idea that I can sleep through anything. But there was something altogether different about the nighttime sounds of Ngwezumba. There was a different quality to it. It may have something to do with the fact that in New York the sounds were coming from the street and I was safe and snug in my fifteenth-floor apartment. Here we were down amid the sounds, and there was always the chance that we would be joining in the festivities at any time, in one role or another.

We stoked up the fire for the night, extinguished the lantern, and turned in. Lying on our cots in the dark just gave our imaginations a lot more scope in interpreting the sounds.

"Hey, Chris."

"Yeah?"

"Mind turning down the volume a bit?"

"I wouldn't mind pulling the plug altogether."

"Right."

We lay for a while longer in the dark. Then I turned on the flashlight and fished around in my footlocker until I found a bottle of scotch. "How about a nightcap, Ted?"

"Best idea I've heard all night."

10

Deacon was uncharacteristically silent on our first morning at Ngwezumba. There was none of his usual humming to himself as he clattered around the cooking area preparing our coffee and breakfast, and not much of the clattering, either. He served us stiffly, without his usual morning greetings, but let us finish eating before confronting me with what was obviously a prepared message from the crew.

"This very bad place, boss. Crew not like it here. Very afraid. Crew want to leave this place, boss."

"What exactly is the crew afraid of, Deacon?"

"Elephant, boss. Elephant most cruel. Crew very afraid de elephant, boss. De elephants gone kill us."

"Deacon, the elephants mean us no harm. We just surprised them last night. We camped in the middle of their trail, so they had to go around our camp and make new trails. It is very noisy, but they won't attack us. They will get used to our being here and then everything will be OK."

This little speech was the distillation of arguments I had spent long hours over in the night devising to convince myself. Although I still had a great deal to learn about elephants, on general grounds I thought that I must be right. We were neither prey nor enemy of elephant, and as long as we were not threatening them I could see no reason for them to attack our camp. After all, people have been living in the bush in Africa for thousands of years under these conditions. While this may sound a bit theoretical, I certainly had no intention of giving up this campsite, and potentially our entire Chobe campaign, just because of a lot of noise in the night. The main problem was to keep the crew from mutinying. I must have said my piece with enough conviction to satisfy Deacon that I knew what I was talking about, because he went off to report what I had said to the crew, and for the moment there was no more grumbling about the campsite. But I knew that my words were half bluster, based more on my instinctual respect for the intelligence of the elephants than any real familiarity with their behavior. I didn't have any more experience with African elephants in the wild than anyone else in camp, with the exception of Thomas, but it wouldn't have been politic to be seen asking his counsel at this juncture.

Teddy and I went out to inspect the woods beyond the camp perimeter. There were a lot of fresh light scars on trees where branches had been broken off, and a few trees had been completely splintered and uprooted, but there was nothing like the utter destruction we had imagined from listening to the pandemonium during the night. The interpretation of sounds heard in the dark can easily be an exaggeration. The elephants had actually been quite economical in their disruption of the forest. They had cleared enough room for just a single file to pass by on each side of us. There was not much sense, I supposed, in their destroying their own food supply.

We worked very cautiously that first day, taking the two Land Rovers in convoy off to the west, retracing our route of the previous day back to the Savuti. We split up there, Teddy going off to the Gubatsa Hills to install an instrument, while I went to the Goha Hills, a group of isolated kopjes off to the north. We rendezvoused back at the Savuti in midafternoon. We had both found very good high-gain rock sites for the instruments and had installed them without incident. In both places we had found it necessary to protect the sites with heavy log and wire enclosures, because we had found the rocks inhabited by troops of baboons. Baboons are very curious creatures with extremely vandalistic tendencies. Anything they find that is strange and curious they are likely to rip to shreds with their sharp teeth and powerful jaws.

The return trip was uneventful. We saw many large herds of animals, including elephant and cape buffalo, but these were at a safe distance and caused us no anxiety. So we reckoned our first day in the Chobe to be a signal success, and we were in much-improved spirits. As far as finding good sites for the instruments, it had been our best day yet. We had had no scares with animals, so I was feeling more optimistic about our chances of settling down to a working routine at Ngwezumba. That night in camp reinforced my conviction about the basic safety of our situation. It was clear that the elephants were becoming accustomed to our presence. We could hear them shuffling by in the dark, but there was much less of the confused barging around and breaking of trees of the night before. The racket from the water hole was as nerve-racking as ever, but it was starting to become a loud and raucus form of background noise. I was no longer imagining phantoms coming out of the night at me from all directions. Nonetheless, there would be no more trips to the woods to satisfy nocturnal calls of nature, the sound of the lions prowling around putting a pretty good block on any such urges. As Uncle Hub from Texas would have said, "We would just have to tighten up our puckerin' strings."

None of this took the edge off the intense alertness that I felt night and day. We were completely on our own in a Nature that was rawer and more potentially dangerous than any I had ever experienced. Here we were intruders, whose presence was tolerated at the sufferance of the animals. As long as we minded our p's and q's we should be all right, but any misstep on our part could have disastrous consequences. If anyone got seriously injured, we would be in big trouble. We would first have to transport the victim to Maun, a day's travel, from where he could be airlifted to the nearest real hospital, in Gaborone. I was responsible for the safety of the entire party, so this meant that I had to take a crash course in the psychology of several species of animals, the most important being elephant.

Two years earlier I had led a similar microearthquake survey in the South Island of New Zealand, during which, in a period of three months our party had transited the length of the Southern Alps to study the Alpine fault, New Zealand's equivalent of the San Andreas fault in California. My team had consisted of three green American graduate students, and we immediately had crises in the field because of our inexperience with the terrain. Fortunately, I had been extremely lucky to have obtained the services, for the first month, of a native New Zealander who had an intimate knowledge of the South Island bush. This man, Ernie Annear, had been born and raised in Fiordland, in the wild and isolated southwestern tip of the island. Starting out as a sheep cockie, he had then spent a few years as a government culler, thinning the herds of red deer that had been imported from Scotland and had since run rampant, destroying the New Zealand forests. He had been discovered one summer many years earlier by a geologist of the New Zealand Geological Survey and signed on as a field technician, in which role he eventually acquired an almost legendary reputation. Rattling around in a Land Rover with Ernie each day, I managed to soak up so much of his lore that after a month of it I was much more in tune with my surroundings than any city-bred New Zealander was likely to be after a lifetime of weekend bush walking. It was this knowledge that, more than any other thing, allowed us to carry off that project successfully and without major mishap.

My present circumstances were similar, though certainly more urgent. I was lucky again to have an excellent tutor in Thomas, who in his own way was ever as much of a landsman here as Ernie was in New Zealand. I was beginning to know him well, and my opinion of him improved steadily day by day. Thomas was the son of an Englishman, hence his name, and a Tsetswana woman, although from his slight, bandy-legged form and wrinkled bronze complexion I imagined he had more than a little Bushman blood in his veins. He was, in fact, a man of some standing in Maun. Beside his job with the U.N.D.P., which was itself a prestigious position for a local man, he owned several cattle posts, which supported about forty Bushmen and their families. He once took me to meet his family at his large kraal on the outskirts of Maun. There I met his four wives, who stood at the doorways of their individual huts bowing deeply to me, while a dozen or so of his children excitedly milled around us. The number of both his wives and offspring were indications of his considerable status within the community.

My initial impression of him had been that he was something of a bullshit artist. When we rode around together each day, he often was taciturn for long periods, but when I got him going on something, he could become quite garrulous and tell some of the most outrageous stories, like how he had hunted elephant with his old shotgun loaded with ball shot. Watching him closely every day, I found his observations and suggestions almost always to be correct and astute, so as time went by I tended to give greater credit to his stories, no matter how outlandish they often seemed at first. There was also his pixie-like way of moving about in the bush, which to a white man's eyes didn't seem properly serious or careful. As I watched him move that way, particularly in the presence of animals, which seemed scarcely to notice him, I gradually realized that this was a peculiarly Bushman way of moving, and that he was in an environment, animals included, in which he was completely at ease. None of the rest of the crew had this manner of moving. If anything, they were clumsier than I in the bush.

My first tutorial on elephants began the next day, when Thomas and I went off to find the Shinamba Hills. These were the mysterious kopjes that were only vaguely located on our maps and that we hadn't been able to spot from the air on our recon flight. They were off to the south of us somewhere in a part of the Chobe in which no roads were marked on the maps. This didn't mean there weren't any—we found plenty of tracks in the sand going off in that general direction. The problem was to find the right one. Thomas approached this as a tracking problem. Each time we came to a place where a new track forked off, he would stop and go study it on foot, to decide which had been more frequently or recently used. Then, climbing back in the Land Rover, he would pronounce, "OK, we follow this spoor," and drive off down the indicated track.

This was a region of well-treed savanna, good elephant country, and we saw plenty of evidence of them. Their spoor littered the tracks, which were frequently blocked by trees they had knocked down. Whenever we came to a small pan or other muddy area, their deep footprints forced us to slow down to a crawl lest we break an axle. In one such place we observed a herd of them crossing the track ahead of us. Thomas continued driving toward them for several hundred yards and then stopped, leaving the engine running. "Dey traveling in herd. Da babies and mommies in the middle. Bulls are outside, guarding. We wait, let dem go by."

We watched as they slowly made their way across the track. They were in no particular hurry. "How close is it safe to approach them?" I asked.

"You get two hundred meters to dem, de bulls dey stop and watch you. One hundred and fifty meters, dey lift da ears and wave at you. That is warning. At one hundred meters, dey charge. I show you." He engaged first gear and drove slowly toward the herd. It was just as he had said. When we got to a point about two hundred meters away, all the bulls on our side of the herd stopped in unison and faced us, watching carefully. He went a little farther and the nearest bull perked up his ears and started flapping them. Then another started doing it. Loud trumpeting relayed some message around the herd. They seemed to be getting fidgety, and so was I. "OK, now we go back," Thomas said, reversing the Land Rover. "This kind of place very dangerous. Too many elephant tracks. We can't go fast. Dey gonna catch us if dey charge."

It took us two days to find the Shinamba Hills. On the first day we ranged all over the eastern part of the search area, following dozens of roads selected by Thomas. At one point we discovered a graded road that ran in a straight line for five miles before petering out in the middle of nowhere. It was just at the edge of the Park, and I guessed that it must have been built for some prospecting project, since abandoned. On the second day I spotted some baobabs, and thinking that they must indicate a submerged rock ridge, told Thomas to take a track out in their direction to see if we could find more. We continued to follow the baobabs, which appeared here and there at infrequent intervals, but which formed a weakly defined line running off to the southwest. After five miles of slow going, we finally spied rock off in the distance within a thicker grove of the trees. They were within a forest in which a herd of elephants was grazing. We stopped to consider the situation.

"Dey grazing, boss. Dey gonna stay dere all day, maybe two days. But dey not moving in herd, like yesterday. Dey spread far apart. Maybe two in one place, three in another. We can try go there. But we must go very slow and not make any loud noise. You gotta be careful de loud noise wid de elephant, boss." He tapped at the horn with his fingernail. "Never use de horn wid de elephant. He think it de trumpet of de strange elephant. He charge you quick."

We stationed one of the crewmen on the roof of the Land Rover as a lookout, and shifting into four-wheel drive, eased our way in among the grazing herd. The elephants were browsing in small groups, which we carefully wound our way between, following the signals of the lookout as he spotted the beasts from his vantage above the brush that blocked our view from inside the cab. Thomas kept the Land Rover lugging along in second gear, trying to keep the pitch of the engine as steady and low as possible. Whenever we came within their view, the great pachyderms would look at us curiously for a few moments, and, evidently deciding that we were not threatening, they would resume their foraging. It was tense going, but we made slow and steady progress, and eventually came up to the rock outcrop where a troop of baboons gathered to watch us. The Shinamba Hills were only fifty feet high at their maximum, and less than half a mile long. They barely poked out of the thin forest canopy, so it was easy to see how we could have missed them from the air. Now that we had found them, it didn't matter how big they were, they were plenty adequate for an instrument site. We installed an instrument that would run unattended for four days. It was a long, slow trip down there from camp, and with all the elephants in the vicinity it was not a site that we wanted to visit every day. I decided not to use the eight-day recording option, though. After our experience in the Kgwebe Hills, I did not want to take the chance of losing that much data because of an instrument malfunction.

We managed to find a fourth rock site not far from camp. That whole general region was overlain by a very thin veneer of sand, within which we found a small region of rock outcrop a few miles northeast of camp, where we installed an instrument. To try to find a similar site for our fifth seismometer, we followed that rocky trend far off to the southwest. If we could find a rock outcrop in that area it would be ideal, because it would be in the center of the array formed by the four units already installed. There was a lot of float in that area, with rough boulders up to ten feet across. This meant there was basement rock close to the surface, but to find if any of it outcropped Teddy and I had to walk the whole area, carefully examining and ringing each boulder with our rock hammers to test if any were actually outcropping pieces of basement rock rather than boulders resting loosely in the sand. Basement outcrop will ring with a distinctly higher pitch than a loose boulder.

We were doing this one day, walking in parallel lines about a quarter mile apart, when Ted beeped me on my walkie-talkie. I picked up, "Yeah, Ted. What's up?"

"Elephants coming through. From the northwest." I looked around and could see the leaders, spread well out. They had cut us off from the Land Rovers where the crews were waiting, parked on the track more than a mile and a half away. Stupid. I hadn't realized we had gotten so far from the vehicles.

"OK. I see them. How many do you figure there are?"

"A good-size herd. I'd say a hundred."

"OK. Well, there's no chance to get back to the vehicles now. Find yourself a big rock, hunker down by it and bide your time. We'll have to let them pass us by."

It was a grazing herd, well spread out, that was slowly drifting through. The feed there was very sparse, so they were moving along in a steady direction as they browsed while strolling. All we could do was sit as quietly and as inconspicuously as possible until they went by. It was tense and tedious to crouch there motionlessly in the intense heat. I was sweating like a pig, attracting swarms of flies that would promenade across the salty incrustations on my face. The flies were maddening, but I could only whisk them off discreetly whenever no elephants were looking my way. Every once in a while one of the elephants would catch a whiff of me and with a snort, wheel away to avoid this evil-smelling creature. Ted and I were pinned down like that for more than two hours before they cleared off enough for us to make it back to the road where the crew was waiting.

We did a lot of waiting around for elephants. On the road to Kasane, that was our new supply base, there was a big mudhole beside the road which was a very popular wallow among the elephants. We would often have to stop there to wait for some elephants to amble their way across the road on their comings and goings to the mudhole. This might take a half-hour or longer, depending on how many there were. We would sit there watching them, all the while keeping one eye on the rearview mirror in case some came up behind us, which could make things a bit dicey if we got trapped between two groups.

Samuel, Teddy's driver, had been well trained by Thomas and handled himself coolly in such situations. Teddy came into camp one day and related an incident from that afternoon when he was out alone with Samuel. "Hey, I got chased by an elephant today, heh, heh. I kid you not. Remember that mudhole on the way to Savuti? The one right by the turnoff to that little round hill off to the north? Well, I spotted a pile of broken-up elephant bones over there. You know how they say that the big bulls go back and break up the bones after the scavengers get done with 'em? It was like that, an elephant graveyard. So I go over to check it out, thinking maybe I could get me some ivory. I reached down to pick up a piece of ivory, and I swear, just as I touched it a young bull pops up out of nowhere and charges me. No shit, it was just like he was guarding it, or something. I took off like a sonofabitch and by the time I hit the road Samuel already had the Land Rover revved up and was rolling off down the road with the door flapping in the breeze. I jumped in just as he hit second and we took off outta there like a scalded dog. I got the ivory, though, heh, heh," he chuckled and held out a six-inch piece of tusk.

Most of the laborers, though, remained terrified of the elephants and whenever we encountered them would cower in the back, rolling their eyes and jabbering at each other. Ted and I were careful to not display our own nervousness. It was important to show by our example that the situation was under control and that there was nothing really to be worried about.

This teaching by example has its limitations, however, when it comes to visceral reactions like fear. I myself am not afraid of most types of animals as long as they are in their natural habitat, but this is a feeling, based on trust and respect for their rational behavior, that is not easily instilled in others. Once I took my wife-to-be back-packing in the High Sierras above Yosemite. She is Japanese and has a typical Japanese reverence for Nature, deep but rather theoretical, so I thought it would be a good idea to show her a little bit of real wilderness. One night we camped in a col at about eleven thousand feet, on a beautiful meadow on the shore of a lake. Yosemite has a large bear population, attracted by the garbage dumps in the valley, and since we weren't more than five miles from the valley rim, I figured that there would be bears prowling around there at night. I took the usual precautions, emptying out our packs and hoisting all our food up a tree. That night the bears showed up as expected. It was a full moon, and the shadows of the bears loomed against the walls of our tent as they passed by just outside, like a lantern show for bears. They nosed through all of our stuff, finding the tins of sardines that I had left in the pack, figuring they couldn't smell them through an unopened tin. They couldn't, but they recognized sardine tins all right, and proceeded to crunch them up just outside our tent. They also found the honey container that I had left out for my evening coffee and then hidden in a mess kit. We listened to them batting that mess kit all over the meadow, but they never did succeed in getting it open. We had no food in the tent, and they didn't bother us there at all, except one that pushed his big snout against the side of the tent and snuffled at us. Yoshiko was very nervous about all this and asked me what we should do. I told her not to worry. "Just go to sleep and snore a bit like a bear, and they won't bother us." Then, to set a good example, I promptly followed my own advice. She was not at all pleased with me the next morning. She had not slept a wink.

I don't know why lions have this popular title "King of the Jungle," because it is certain that the elephants rule. We were not really concerned about lions, despite listening to their lusty roaring every night. They would be a great danger to anyone wandering around the bush at night, of course, but we certainly had no intention of doing that. Cape buffalo, on the other hand, presented a different sort of hazard. They are enormously powerful beasts and can do an awful lot of damage with their terrible sweep of horns. They are about as intelligent as their domestic cattle cousins but much more massive, and they have extremely malevolent dispositions. We would often see them in herds of several hundred, grazing in the grasslands or lurking under the trees. The only sound advice about dealing with them is to stay well clear. They have poor eyesight: They cannot see much at a distance and not much even at short range unless it is moving. So whenever we had to pass a herd of buffs, we would ease by, trying to maintain a nice steady speed, and watching them carefully all the while. As soon as one of them would notice us, they would all turn and stare at us with their beady-eyed gazes. They have a strong tendency to stampede. This is an unstable form of herd behavior, and difficult to predict, which is what makes it especially dangerous. Once one moved toward you, for whatever lamebrained reason, the others would too. That made the first one move faster, and so on, and before you hardly notice what is happening they all would be stampeding full speed directly at you. This is a most terrifying thing to witness, because once they start stampeding, there is nothing to stop them from trampling to death anything in their path. The main strategy with buffalo was to not do anything to start them stampeding, but always to have a well-planned escape route in mind just in case they did anyway.

By the time we had been in the Chobe for a week we were doing our work on a more or less routine basis. Everyone was in a high state of alert, paying careful attention to the animals, but this was becoming a more normal part of our existence. The psychological state of the crew seemed to be stable, and they were working well now. Somehow, probably on one of our visits to Kasane, they had heard about the mutiny of the construction crew that had been at Ngwezumba before us, and that story seemed to confirm their own fears and play on their anxieties. The sight of the unfinished bridge was a daily reminder of this. There was still a lot of tension among them, and I was concerned that any serious incident would be likely to put them back into a mutinous mood.

Eventually, such an incident did occur. We had been out on one of our usual runs to the west, tending the Gubatsa and Goha Hills instruments as well as doing some exploration a bit further afield to see if we could trace the line of kopjes farther to the north, where we wanted to shift one of the stations. It was a long circuit, but it was made much longer by the extraordinary number of flat tires we had that day. Flats are common because of the thorns that get caught in the tires and gradually work their way through. As a result, we always carried two spare tires in each Land Rover. If we had any more flats than that, a tire would have to be repaired on the spot, which could produce lengthy delays. On that day, I think we had had something like seven flats between us, which was why we were so late for our rendezvous at the Savuti.

We took stock of our situation. It would be touch and go if we could make it back to camp before nightfall. Being out of camp, and especially traveling after dark, was about the most dangerous thing you could do in the Chobe. It was the number one taboo. We had nearly a hundred miles to go to camp, and no more than two hours of daylight left. It was really a question of what speed we could do on the rough roads, if pushed. That was something we could only guess at, since normally we traveled at a sedate pace, being mindful of the wear and tear on the vehicles. Our only other choice was to backtrack to the Savuti game scout camp and spend the night with them. The problem with that option was that we had no way to contact our camp to tell them what we were doing, and if we didn't arrive back that night it would be likely to put them into an absolute panic. I wanted to avoid that if at all possible. They were in a nervous enough state already. After debating this problem for a few minutes, we opted to make the run for camp.

I was riding with Samuel in the lead car, with Teddy following behind in the other with Thomas. We sped over the flats, coaxing sixty-five out of the Land Rovers as we rocked along, barely in control. Land Rovers are not built for speed, but no machine would have been able to do much better on those rough roads. Eventually we realized that in spite of these sprints we weren't going to make it in time. There were too many dry stream beds and soft sandy spots to cross, which slowed us down so we weren't averaging much more than forty miles an hour. But by then we had passed the point of no return and had no choice but to continue.

It happened less than twenty miles from camp. We were driving, in spite of the dust, closely bunched together for safety. It was at that interval of dusk when you need headlights but it isn't yet dark enough for them to do much good. We came booming around a corner and down a little hill and saw, in the half light, a herd of elephants spread out across the road in front of us. There was no way to avoid them. Jamming on the brakes, we went skidding, all wheels locked, right into them. Miraculously, we didn't hit any of them, but we ended up stopped in a cloud of dust right in the center of the herd. Samuel switched on every available light and kept his foot on the brakes to keep the brake lights on, which lit up the swirls of dust in a hellish red glow. I felt like a little kid caught in a crowded elevator during an earthquake. Everywhere I looked I saw huge knees milling about. Samuel revved the engine to make it roar loudly. He didn't touch the horn. We needed to let them know where we were, so that they could avoid us, and we needed to scare and mystify them just enough so they wouldn't attack. The crew in back were lying on the floor with their arms over their heads, moaning piteously. I stared around in wondrous terror, as the elephants milled around us, cows pushing terrified calves past the car, not a one of them touching us.

After an eternity that could not have been more than a minute or two, we just as suddenly found ourselves alone. The bulls had regrouped the herd thirty or forty yards across the road, and when finished with that task, they turned to see what demons had attacked them When the dust settled, they recognized us for what puny enemies we were, and with a blare of trumpeting, they charged. Samuel and Thomas let out the clutches and we got out of there, fast.

We crept the rest of the way in the dark back to camp at a sober, deliberate speed. It was nearly ten o'clock when, with great relief, we finally spied the big bonfire on the bluff and roared up the side of the ravine into camp, where we were greeted by the rest of the crew, who had been awaiting us with great anxiety.

11

The following day did not pass without repercussions from our encounter with the elephants. We stayed in camp until midmorning because it was a day scheduled for going in to Kasane for supplies. I was first approached by Blackie, who had a roundabout way of discussing the problem.

"Boss," he said, "We got to be leaving this place soon. Da crew, they want to go home. Soon the rains come. When de rains come, their houses going to melt. Dey got to be home to take care of them."

"I understand, Blackie. But now we are getting good data, so we have to stay a while longer. There is no sign of the rains yet. Don't worry. When the rains come, the crew will be allowed to go home right away."

"OK, we stay for a little longer, boss. But remember, this crew is volunteer. Dey already did the field time this year. When dey want to go, dey can go."

I knew that this business about the rains was Blackie's way of avoiding the root subject that he didn't want to mention, the crew's fear of staying in the Chobe. The coming of the rains would be preceded by a change in the cloud patterns, which would provide a warning of a week or two, giving us plenty of time to wrap things up and return south. There had been no sign yet of this change in the weather, and he well knew it. On the other hand, his other remark was meant as a warning to me. This crew was indeed a volunteer one. All of them had previously fulfilled their contractual obligations for annual field time. They had signed on for this trip, rather than spend the last months of the year doing chores around headquarters in Lobatse, for the small bonuses that they would receive and for the fun of seeing a new part of their country. This wasn't turning out to be as much fun as they had expected. I knew

that if they all agreed under Blackie's leadership to return to Lobatse, there would be nothing I could do to prevent it, nor would
they face any punishment from the Geological Survey. It was up to
me to hold it all together, but that wasn't going to happen if there
were any more incidents with elephants. They were a loyal crew,
but they could be pushed too far.

This wasn't the last word on the subject for the day. Teddy and I
were having a drink that evening before dinner, when Deacon
stormed over, and pointing accusingly at the whiskey bottle, said,
"That is why you brave, boss," then stalked away.

"Huh?" I looked at Teddy. He looked back, just as dumbfounded,
then grinned. "Heh-heh-heh. Maybe he's right."

Deacon certainly had made a good shot, even though it was a bit
wide of the mark. The whiskey was no doubt a good aid for sleeping through the raucous nights, but I don't think it does much to
enhance bravery or courage, unless one mistakes that for foolhardi-
ness. As far as courage went, I tended to share the opinion of Saint-
Exupéry: "It's not made up of very pretty feelings: a touch of rage, a
tinge of vanity, a lot of stubbornness, and a vulgar sportive thrill."
As far as I was concerned there was more than a lot of stubbornness
involved. We were out there to do a job, and damn it all if we
weren't going to get it done. You can't be very successful as a scientist without a heavy dose of stubbornness, or determination, whatever you want to call it. It's a competitive business in which you
must be able to solve problems that others, sometimes many others,
have failed to solve, or haven't even seen in the first place. To be
successful at that game time after time requires more than good scientific skills and a gift of cleverness. It also means having the will to
not give up where others might throw in the towel. Add a tinge of
vanity for a little ego motivation, and there you have the right ingredients for a good scientist.

Inexperienced as Ted and I were to this life in the African bush,
we were less likely than most to cave in and give up in the face of
events. We may have been tinhorns from New York, but there was a
little more to it than that. For me, studying earthquakes was not a
mere profession, but more of a calling, and one that I had been
called to early. My grandfather on my mother's side had been a geologist. I had never known him except from some yellowing pic-

tures of a stout, strong-featured man with a big mustache and an impressive name, Washington Henry Ochsner, standing in stilted poses in field camps, wearing the field clothes of an earlier era, a wide flat-brimmed hat with boots laced up to the knees over baggy trousers. He had been the chief geologist on the voyage of the *Academy*, a three-masted schooner sent to the Galapagos in 1905-7 by the California Academy of Sciences to survey the geology and biology of the islands and to investigate some of Darwin's findings from his visit, more than a half century earlier, on the *Beagle*. As a result, he had missed the great San Francisco earthquake. It must have been something to sail back through the Golden Gate after a two-year voyage and find the city gone. I had stories only from my paternal grandmother about the earthquake. Her main recollections were of the fire and thestreets alivewith fleeing rats.

My grandfather soon left academic work and formed a company engaged in the early days of oil exploration in California. He died young, many years before my birth, of tuberculosis contracted in his exploration camp in the Kettleman Hills, and a few months before the oil finally came in at the North Dome. This was all part of family legend, inculcated in me at a young age by my mother and predisposing in me an early interest in geology. But it was another event that focused me much more clearly on what would become my life's work.

In 1952, when I was nine years old, a very large earthquake occurred in Kern County, in Southern California. I was standing in the front yard of our home in the San Ramon Valley, east of San Francisco Bay, when the long train of surface waves came through, with first their rolling and then their slow, swaying, side to side movement. Wow! My mother appeared at the front door and shouted at me to come. I rushed up, breathless with excitement, "What was that, mom?"

"It's an earthquake. Quick, get inside the doorway." She was obviously terrified, which impressed me even more.

"What's an earthquake?" I asked, awestruck.

"Ask your father when he gets home," she said, regaining her composure as the earth settled back down to its usual stability.

I waited impatiently all day for my father to return and explain this enormous event to me. When he finally arrived, the first thing he said was, "Hey, did you feel the earthquake?" I was astounded. My father worked in Livermore, a long way from where we lived. How could he have been shaken too? He had the paper with him, and it had pictures of a town that had been completely wrecked by the earthquake. The town was called Tehachapi, and it was three hundred miles from where we lived. I was beginning to get an idea of the tremendous power of this thing.

My father kept putting off explaining to me what an earthquake was, finally promising to do so after dinner. When I eventually cornered him, he sat down and with the voice he used to explain serious things to me, proceeded to tell that there were big caves inside the earth, and once in a while a big rock would come loose, and falling in the cave, would cause an earthquake.

I was stunned by the absolute absurdity of this explanation. I had grasped the enormity of the energy released by this thing. It had sent out waves in all directions which, at a distance of over three hundred miles, had shaken us very strongly. No rock falling loosely in the earth could generate even a tiny fraction of that energy, nor could there be any cave big enough to contain it. If I had been able at that age to express such a sentiment, it would have been: "Rubbish!" I was truly shocked. This was an epiphany for me: Here was something so big and important, and my father had not the slightest idea what it was or what caused it. From that moment on I was determined to answer my own questions and to find out for myself how earthquakes worked. It wasn't till much later that I found out that my father hadn't actually made up the theory about rocks falling in caves, he was just a little out of date. That had been the idea espoused by Lucretius in *De Rerum Natura* in the first century B.C. This ignorance was not confined to my father. I had found, by the time I got to college, that the modern theory of earthquakes wasn't in all that much better shape. There was a lot of work to be done. I had picked a good research program, a fresh and untrammeled field.

Ted, in his own way, was also destined for the work we were doing. His father had been a chief petty officer in the old Navy. He had served in the river boats in China during the '30s, and Ted could recount stories from his father that sounded like they came right out of *The Sand Pebbles*. When Teddy joined the Navy in the modern era, it must have been disappointing by comparison, and he didn't stick it for long. Life on the *Vema*, the beat-up old converted schooner that for many years was Lamont's primary deep-sea research vessel, was undoubtedly more congenial to his disposition, but it was here in the Kalahari that he was finding the sort of fulfilling life for which he was constitutionally suited. That was why his spirits never flagged, no matter how difficult things might get.

Our new supply base, Kasane, was a very pleasant village on the Chobe River just outside the northern edge of the Park boundary. It was much smaller than Maun and much prettier, full of flowering shrubs and trees, dominated by jacarandas with their lilac-colored blossoms, flowering acacias, and a kind of tree bright with spectacular fiery orange blooms. It was situated a few miles upstream of the confluence of the Chobe and Zambezi rivers, where four countries joined. Zambia was on the north bank of the Zambezi, Rhodesia butted in from the east, and the swampland between the two rivers was part of the Caprivi Strip, the long east-west salient of Namibia. Except for its air safari business, Kasane was a commercial backwater, being sealed off to the south by the impassible Chobe National Park, and being connected to Kazungula on the Zambian side by only a primitive cable ferry across the Zambezi. This was a historic effect of it having been bypassed by the Rhodesian Railway, which crossed into Zambia at Victoria Falls thirty miles to the east in Rhodesia. At the time we were there the railway was closed at Vic Falls in retaliation for Zambia's support for the guerrillas in the Rhodesian war, although resulting events were afoot, which we were as yet unaware of, which would soon change Kasane's situation drastically.

On our first visit we had to stop by the district commissioner's office to get permission to use our fuel chits at the local government depot, because we had only been cleared to do that in Maun. The D.C. was a young man, a hereditary subchief of the local tribe, very handsome and regal in bearing. He conducted his business with the noblesse oblige of a chief rather than the pettifoggery of a bureaucrat, signing our chits with a swift flourish and welcoming us to the district.

The village itself didn't have much to offer. The only shops were a couple of open sheds with little in the way of goods. We would be able to get only the most basic of stores in this place, like meat and mealies, and if we were lucky, some beer. While the crew stocked up on fuel, water, and provisions, Ted and I drove out on the road along the river where tourists are taken to see the wildlife. There were a lot of antelope of various kinds, baboons, and even a few semi-tame lions. While we were there we met a party of tourists riding in a row of Volkswagen minibuses all painted in jungly colors, who gawked and went clackety-clack with their cameras as the guide gave them his spiel through a microphone.

There was no pub in Kasane on the order of Riley's, but there was a tourist hotel along the river where we could get a decent lunch when we were in town. I remember getting an egg salad sandwich on white bread there, which at the time struck me as the most exotic and delicate of luxuries.

We were obliged to pass through a gate on reentering the park. The Main Gate was a much more impressive affair than the South Gate through which we had originally entered the park, with a striped pole across the road attended by nattily uniformed game scouts. Across from it was a small building with a sign over the door: "Richard Dziekowsky, Chief Ranger." I thought it would be a good idea to pay the ranger a courtesy call, so I strolled over. There were a couple of game scouts in the small room, which was devoid of furniture except for a few folding chairs and a desk behind which sat the ranger, a stocky man of fifty or so, with a big and florid, rather brutish face and thinning blond hair. I introduced myself, explained our business, and informed him that we would be camping at Ngwezumba for the next few weeks.

His face turned a brighter red. "You're bluidy what?" he shouted in a thick Afrikaner accent. "You'll be doing no such thing. Camping in the park is forbidden. You go pick up your camp and be out of there by sundown today, hey?"

"You don't understand. This is a government project. We have permission to camp in the park."

"Ja, well you don't have my permission, see, and this is my park. Listen, boyo, Ngwezumba is a bluidy dangerous place. Last damn government party found out all about that. You Mickeys will get into trouble and then it'll be me having to come clean up your nappies. I won't have it, hear!"

Trying to control myself, I managed to say, coldly, "I suggest, then, that you get in touch with your ministry and talk to them about it," and stalked out, letting the screen door bang behind me.

When I got back to the trucks, Teddy said, "What the hell was that all about?"

I climbed in and slammed the door, then took a deep breath and said, "Let's just say that I don't think the ranger and I are going to be pals, okay? Now let's get out of here."

When we returned to camp the following afternoon we found the camp crew standing in a group waiting for us. Blackie angrily explained, "De ranger he come to da camp, boss. He look at everything. He look in the tents. He look at all our stuff. He take Samuel's gun, he take de horns."

They were very upset. I could understand their feelings. The ranger's search was a major violation of our privacy. I found a note addressed to me pinned on a tree. On it were listed a whole series of infractions of minor and obscure park rules: camping at a site not authorized by the ranger, camping within one thousand yards of a water hole, etc., etc. Then there were two items: "Possession of unauthorized firearm. Weapon confiscated. Illegal possession of government trophies. Confiscated."

The weapon was Samuel's shotgun. He was in hysterics over it having been taken. It was a major piece of personal gear for him; his pride and joy. "Don't worry," I said, "We'll get it back."

Ted was equally indignant over his "government trophies" being confiscated. He had found a beautiful set of kudu horns, and another of impala, which he had spent a week of evenings cleaning up and polishing. He had been looking forward to hanging them up in his den at home as a souvenir. They had also confiscated a fist-sized chunk of elephant tusk, which he had already started carving with elaborate scrimshaw. Luckily, it wasn't the piece he had risked his neck retrieving from the mudhole. That one he found in its place stowed deeply in his kit. "What is this government trophy crap?" he complained. "It's not like I killed an animal for them or something. These horns are lying all over the place out here. Nobody's ever gonna come out and collect 'em."

Blackie said, "I know dis Ranger Dick fella, boss. He very bad man. He use to work at de Kruger Park. Dey fire him for abusing de game scouts. Dey say he used the whip on dem."

The next morning I made a formal complaint via the radio-telephone to the Geological Survey. I didn't want to put up with any more petty interference from this man.

Meanwhile, the work was going splendidly. We finally had a network of seismometers operating at high sensitivity, and it was starting to pay off with results. It was exciting to visit the instruments now, and to see on the daily records two or three earthquakes recorded, most of which were large enough that they would be picked up on all the stations of our array. Now, when I sat down each afternoon at the little folding table in the instrument tent that served as my office, I had more to do than just make a daily entry in my journal. Laying out the daily haul of drum recordings, I would indentify the wiggles that marked each tiny earthquake and then measure off the arrival times of the seismic waves at each station, which would allow me to pinpoint the event. This was made easy because Texas Instruments had just come out with the first programmable pocket calculator, and I had written a quick and dirty earthquake location program which could be read into the calculator with a little magnetic card. All I had to do was enter the position of the instruments and the arrival times of the seismic waves, and it would almost immediately read out the latitude and longitude of the earthquake on its little LED display. It was a lot quicker and more accurate than the old field method in which the locations were done graphically by laying out lines on a map and finding their points of intersection.

As the days progressed, the circles marking the earthquake locations on my map were beginning to form a pattern. We were picking up events as far south as Maun and as far east as the Zambezi gorges in Rhodesia. Many ot them were starting to define a swath running southwest from the vicinity of Kasane down through the Mababe Depression toward Maun. This was consistent with their occurring on a set of faults striking off to the southwest that lined up with the lineaments that I had seen on the satellite images. It looked as though my working hypothesis was being confirmed, although final confirmation would have to await the determination of the slip directions of the earthquakes. This required computations beyond the capability of my little TI calculator and would have to be done back in the lab on a proper computer.

I could do a rough graphical solution for the fault motions, though. The idea behind this is very simple. An earthquake arises when the two sides of a fault slip past one another in a rapid shearing motion. The sign of the outward motions in the Earth caused by this disturbance are not the same in all directions. The first motion of the first seismic wave to reach an observer will be compressional if the earthquake motion is toward the observer, and dilational if the motion is away from the observer. To visualize this, consider the fault plane as being horizontal and that during the earthquake the upper block slips from left to right relative to the lower (both blocks actually move). Imagine a vertical line through the center of the figure, dividing it into quadrants. Then an observer in the upper right and lower left quadrants will receive a compressional first motion, because the earthquake motions in those quadrants are outward "pushes." Similarly, a dilitational motion will be received in the upper left and lower right quadrants, where the earthquake motions are inward "pulls." On the seismogram, compression is up on the record, and dilation is down, and I had been routinely recording the polarity of these first motions on the seismograms. Now that I had located enough microearthquakes, I could work out the directions of the ray paths traveling from the earthquakes to each of the recording stations. Plotting the ray directions and polarities on a stereographic projection, I could graphically work out the planes that divided the four quadrants, which would tell me the rough orientation of the fault plane and the direction of slip on it. When I did this, I could tell already that the orientations and slip directions of the faults pretty much corresponded with those predicted by my hypothesis that the faults were creating a rift valley.

We were also beginning to get a decent measurement of the rate of seismic activity of this region. In the south we had been stymied by having to put most of our stations on sand and so had recorded too few events to get a good estimation of the true level of activity. Now the data were beginning to show a fairly high rate of activity, comparable to that of the East African Rifts in the north. There had been previous microearthquake surveys done in northern Kenya, the Ethiopian Highlands, and in the Ruwenzori Mountains of

Uganda in the Western Rift, and the rate of activity measured in those places was about the same as we were now observing in Botswana. This fact was also consistent with my hypothesis that this region concealed an active rift segment.

The fieldwork was also going very smoothly. It was a great pleasure to go out every day and drive through this area teeming with wildlife. Each day we would pass vast mixed herds of antelope. I could gaze at them for hours without tiring of the scene. From time to time I would call a halt and try to photograph a particularly beautiful scene with my long lens. Elegant sable antelope, grand stately eland, giraffe grazing on the treetops, wildebeest charging across the veld in one of their mad stampedes, a family of warthogs scuttling through the underbrush, each in their turn would fill my viewfinder with their beauty. I was no great wildlife photographer, but you didn't need any special talent to take good pictures in this place.

The zebra had a peculiar game they liked to play with us. Whenever we passed a herd of them anywhere near the road they would break into a gallop and race along next to us, twenty or thirty strong. After a minute or two of this, there would always come a point when the leader would get just far enough ahead of the Land Rover that it would dash across the road in front of us, followed close on its heels by the rest of the herd. We would always have to jam on the brakes to avoid hitting the laggards and would end up stopped in a cloud of their dust. As soon as they all got across they would stop and wheel about to face us, looking at us as if to say: "Hah, we won!"

We saw many other curiosities. Teddy and I were changing a record one day at a site in a grove of trees with beechlike leaves. Teddy let out a chuckle and said, "Look at this, Chris," pointing into the lid of the instrument case. I looked and all I could see were some dried up old beech leaves. Teddy poked one of them and it started walking. We picked it up and looked closely at it. It was an insect, whose body resembled the center stalk of a leaf, and from this protruded some ragged flat sections so that the whole effect was of a dried up broken beech leaf. What a specialized feat of camouflage this was; I supposed that this insect could live only under beech trees. Later I saw an example of the opposite, an insect with a completely general method of camouflage. I spied this creature at another instrument site, this one in a sandy place. As I was kneeling by the instrument, I thought I saw a bit of sand moving. Looking closer, I saw that it was a beetle whose back was completely covered by sand. On examination I found that this beetle exuded a sticky substance on its back, so that it soon became coated with whatever soil was about, making a camouflage perfectly tailored for whatever surroundings it happened to be in.

We often set up our camp chairs on the edge of the bluff in late afternoon and watched the spectacle of the daily animal migration to the water hole. On one such occasion Teddy remarked, "Gee, it must have been just like this a couple of million years ago when Lucy and her tribe were roaming around in this country." We had only just heard on the BBC of the discovery, in Ethiopia, of the nearly complete skeleton of a female hominid, nicknamed Lucy.

"Yeah, it probably is, but I doubt that Lucy's crowd ever spent much time out in this kind of country."

"No?"

"No. Think about it. The early hominids had only the most primitive stone weapons. They would have been no match for lions, and since they were very slow compared to antelope, they would have been among the easiest prey out in the open savanna. That's not what you want to be if you want to promulgate your species. They must have lived in more protective habitats, like the apes, who live in the rocks or in trees deep in the forest."

"So where's that?"

"The Rift. I'm no anthropologist, but I don't think it's a coincidence that all the finds of early hominids have been in the Rift. I don't think that is just where they are preserved either. Geof King, who sometimes works on the relations between tectonic activity and human settlement, thinks the Rift was their main habitat, and I tend to agree. The Olduvai, where so many of the finds have been made, is a gorge leading off the edge of the Serengeti plain into the Rift Valley. A few million years ago it would have been much closer to the Rift floor. Geof argues that it would have offered plenty of safe refuges with its steep terrain, and still have easy access to the plain for hunting forays. Lucy was found in the Awash Valley, in the Afar. In fact, a suspicious number of the Australopithecus sites are in the volcanic parts of the Rift, in volcanic ashes intercalated with lavas. That would have been a perfect protective terrain for early man. Fresh lava flows are virtually impassible to large animals, but they can immediately abut areas with abundant food sources. Modern man, on the other hand, who can cope with the animals, avoids living in such inhospitable places."

"Another thing that bothers anthropologists is the rapid dispersion of early man. They have discovered hominid sites of almost the same age on opposite ends of the continent. What would cause such massive migrations? Modern cases like the Bantu expansion were caused by population pressure, but there was never a large population of early man. I don't think they ever actually dispersed over the continent. I think they went up and down the Rift; it was their habitat corridor. Their continent was the Rift, sunk in the larger continent. If any place deserves to be called the Garden of Eden, it's the African Rift Valley."

Riding around with Thomas every day, I picked up a lot more lore, as well as a few tall tales. I was particularly interested in why the crew showed no fear about the kopjes we were visiting here in the Chobe.

"Dese not Bushman kopjes, boss. Not like Kgwebe Hills or Tsodilo Hills. Bushman never come here. Too many elephant. Elephant too strong for Bushman."

"Oh, but I thought the Bushmen hunted elephants?"

"In de old days, it true. Many Bushmen shooting da little arrows with da poison at de elephant. It true, I saw de pictures on de rock. But now de Bushman, he very weak. He don't have de power like de old ones. Now dey afraid de elephant."

I also got him to open up a bit more about the Bushman spirits in the kopjes. He told me that if the Bushman put the curse on you, your spirit was locked inside the kopjes, where it stayed a prisoner forever.

"It true, boss. I heard the voices of de spirits in the kopje many times. Many times. Sometime you camp near the kopje, you can hear the spirits talking from inside de stone. Sometimes at night, in the storm, you can see de spirits walking on the top, carrying dere lights."

Echoes, I thought. That's their explanation of echoes. Out there in the desert there is no other place to hear echoes except from kopjes, so it is easy to see how the two would be associated. And the lights in the storm must be some kind of electrical phenomenon. I was entranced.

He told me many other stories about Bushman superstitions and the kopjes, many of which were complex and not as easy to explain with natural phenomena. Listening to them, I could feel the spirit of generations of campfires on the veld. They were like ghost stories told around the campfire. All good fun. Provided, of course, that none of it was true.

Working out of our solitary camp in the middle of the Chobe, we were beginning to feel a great attachment to the wild veld, but from time to time we had to go into Kasane for supplies and reconnect with human society. On our second visit we ran into a small group of Americans at the hotel, where we went for lunch. They turned out to be pilots, more ex-Vietnam types, who were traveling through Africa hoping to find flying jobs somewhere. We told them about the Air Alaska outfit in Lobatse. At first they were very interested, but after I told them a few details they noticeably cooled. They were ex-Navy types, and it seemed that they didn't expect to get much of a welcome from the ex-C.I.A. guys running Air Alaska. I guess it was a different brotherhood, or something.

When I left the hotel, Teddy was already in the cab of the pickup. He was talking to a black woman who was standing next to the truck. As soon as I got in, he put it in reverse and floored it out of the parking lot, leaving the woman standing there in a cloud of dust. Peering through the windshield, I recognized her: it was Teddy's Number One Girl! "Hey Ted, it's the Groovy Queen from Riley's. How the hell did she get here?"

"The hell if I know," he muttered, spinning the pickup around and gunning it down the road into town.

"Well what is she doing up here?"

Teddy replied in a wincing tone of voice, "She said she came to find me. She wants to come live in our camp with me."

I erupted into peals of laughter. Teddy didn't seem to think it was so funny, though.

We met up with Blackie and the rest of the crew and headed back to camp. We stopped at the park gate where I had to have a confrontation with Ranger Dick over Samuel's shotgun. He gave me this big lecture about how you could have a rifle in the park for protection but it had to be specially authorized. Any other weapon was subject to confiscation and would not be returned. "But look," I said, "it's the man's personal property. He brought it along to shoot guinea fowl for the pot. We've been out here for a couple of months, and there wasn't any chance for him to return it to Lobatse before we came into the park, even if we'd known about the rule." He refused to listen to my arguments, and we got into a bit of a shouting match. In the end I threatened to bring it up with the ministry and he relented, saying we could pick it up at his office the day we left the park.

About a half mile past the park gate a turnoff to the right led to a posh new hotel that had been built for the air safari trade. Its main claim to fame was that Liz Taylor and Richard Burton had been married there on their second time around. We pulled in because I wanted to see if we could get some soft drinks there, since there had been none in town. The place was completely empty because there was no tour in just then and those kind of places don't get walk-ins. We all piled out and went through the fancy lobby into the bar, where the barman served us Pepsis while his assistant went to look in the storeroom for a case we could take back to camp.

We were standing around enjoying our cold drinks when into the lobby strode Ranger Dick, followed by a bunch of his game scouts. He stormed over and waving his fist at me, yelled, "Get these kaffirs out of here! They don't belong in this hotel."

Kaffirs? I flipped out. "What the hell do you think you're talking about? This isn't South Africa, you know. These guys can go anywhere they want." We faced each other, five feet apart. My crew circled around me. His game scouts gathered around him. The barmen stood stock still. I was livid with rage, and about to do something stupid. Teddy put his hand on my shoulder. "Cool it, Scholz." I relaxed just a bit, and caught a glimpse of Blackie's face. He looked like he was liable to clobber the son of a bitch at any time. If that happened we really would be in the soup. "OK you guys," I said, "Grab the Pepsis. Let's get out of here."

12

The national election day was coming up and the crew had to go into Maun to vote, so Teddy and I decided to take a holiday and go to Victoria Falls. My original plan was to drive to Kasane, and crossing the border there to take the road along the Zambezi to Victoria Falls. When I told this to Blackie he said that the best way to go to Victoria Falls was instead by way of Panda-ma-tenga. I had my doubts about that. Panda-ma-tenga was on the Rhodesian border due east of us. Although it was true that on the map it looked like a shorter and more direct route, getting to Panda-ma-tenga meant taking the disused road that we were camped on, and I wasn't too sure about the wisdom of that. In the end, against my better judgment I decided to go the way that Blackie had recommended. I'm not exactly sure what my reasons were for following his advice. Perhaps it was because Blackie was so insistent about it, and I didn't want to hurt his feelings. Or maybe it was just because I wanted to see another part of the country that we hadn't visited before. Then again, maybe it was because Blackie told us that the river road was under constant shellfire attack from guerrillas based across the border in Zambia.

Victoria Falls was about 120 miles from Ngwezumba as the crow flies. But since we weren't crows, we wisely decided to get an early start. Selecting the better of the two Land Rovers, we loaded it up with our rucksacks, three spare tires, and a couple of extra jerry cans of petrol and water, and set out early in the morning. The road heading east out of camp hadn't been used for years and was heavily grown over with grass. It was not so bad that you couldn't see where the road was, but enough so that you couldn't see the holes in it, which we were constantly crashing into with bone-jarring thumps. After a half hour or so of this, Teddy made the obvious comment, "Nice road."

To make matters worse, the road was going through treed savanna, much more thickly wooded than anywhere else we had
been, and it was heavily populated with elephants. Since the road
had not been maintained for years, every few hundred yards we
would encounter a tree knocked down across the road and would
have to make an engine whining detour around it through the thick
brush. And of course, whenever we encountered elephants walking
along in the road we would have to wait politely until they had
wandered off it before we could proceed past. In the first hour we
had seriously considered turning back and taking the route through
Kasane, but that would have meant returning to camp and seeing
the crew again, since they wouldn't have yet departed for Maun.
So, deterred by pride, we muddled on, each mile taking us further
from a decision to abandon this ill-conceived route.

We had many reasons for wanting to visit Victoria Falls. Of
course we had the usual touristic motivations: We wanted to see the
falls and to send postcards to all our friends back home to make
them jealous that we had visited this famous place. But these were
mere diversions compared to the deeper motives that were really
driving our pilgrimage. A few weeks earlier Teddy had found a Vic
Falls newspaper in Kasane, and in it was an advertisement for
Wimpy Burgers. This had started Teddy off on a fixation on hamburgers. I tried to dampen his ardor by telling him that Wimpy's
was a lousy British chain and that their burgers were limp imita-
tions compared to the American version. This didn't work at all. He
couldn't stop himself from talking about hamburgers all day long.
He went on and on about such esoterica as why A&W's Super-pa-
paburger is superior to a Whopper, etc., etc. Then he got off on
milkshakes and fries. I discovered that this kind of food fixation is
highly contagious. It was not long before I started to have my own
visions of the queen mother of all cheeseburgers, with a sesame
seed bun piled high with lettuce, tomato, and onions, and with a
thick juicy meat patty oozing underneath. I could imagine biting
into it and the sloppy sauce mixed with the meat juices gushing into
my mouth. Uhnnn! Man, I got so I could write x-rated ad copy for
hamburgers in my head. I also had visions of an idealized gin and
tonic. This gin and tonic was not like any I'd seen for months. It
had everything in it, like gin, and tonic, and lots of ice, and even a
slice of lemon or lime perched on top. I imagined it in a tall glass,

frosted on the outside so that when you touched it you left wet marks on the glass and made the ice tinkle inside. And when you looked down into it, it had that vaguely blue color that real gin and tonics are supposed to have. All of these wonders could be had for the asking in Victoria Falls. We were exhibiting the classic symptoms of field-goofiness: a psychological reaction resulting from relaxation of the tension built up over an extended period of intense fieldwork.

For hours we sustained each other with these visions on the long road to Panda-ma-tenga. The conditions of the road did not allow us to make more than about ten miles per hour. We had already been on the road for more than six hours and we were still deep in the featureless thick forest. Besides being hot and sticky and generally miserable, we were beginning to get worried. Two of our three spare tires had already been changed for flats, and since we were driving almost continuously in low gear in four wheel drive, our fuel consumption was getting dangerously high. If we didn't break out of this bush soon, we would be in real trouble. By dead reckoning I thought that we should have arrived in Panda-ma-tenga at least an hour earlier, but the constant detours made it impossible to correctly estimate our progress.

Just as we were becoming really desperate, the forest in front of us became lighter and we suddenly burst out into a cleared area. I stopped the Land Rover in astonishment. We were in an absolutely cleared-out area two hundred or three hundred yards wide that ran straight as an arrow off both to the left and right as far as the eye could see. In the middle of this cleared strip was an elevated structure, like a levee. I drove up on it. It was flat on top, graded, and as wide as a four-lane highway. We were both a little giddy from this unexpected and bizarre sight. I asked Teddy, "What in the world do you think this is?"

"If I told you, Scholz, you'd never believe me."

We rumbled down off the levee and picked up our road on the other side. A couple of hundred yards farther on we reached Panda-ma-tenga. It was deserted except for a couple of naked kids hanging around a smoldering fire, some chickens, and a dog. The only substantial structure was a large adobe building that looked like the Alamo after Santa Ana had got through with it. Its roof was fallen in and it was pockmarked with bullet holes. The dirt yard in front of it was full of shell pits. We continued on down the road to the border crossing.

The border was at the bottom of a valley. It was marked by two lines of chain-link fence, fifteen feet high with razor wire on top, that ran off as far as you could see in either direction. Between the two fences there was twenty yards of freshly plowed ground. What a wonderful sight, I thought. My first death strip. I took a picture of it. We talked the guards into letting us through to visit the border post a half mile farther along on the Rhodesian side. The border marked a pronounced change in the countryside. Whereas the Botswana side was flat monotonous bush, the Rhodesian side was hilly and planted with neatly tended cornfields.

The officer in charge of the border post was an Indian, a Sikh, with bushy beard and turban. He seemed surprised to see us. They probably didn't get many visitors in this remote post. We explained the purpose of our visit to Rhodesia and showed him our passports. "I'm velly sorry, sir, but this post is not for the tourist crossing. We only take the local commercial traffic here, sir. By the special permit only."

We pleaded with him, telling him how far we had come, and that it would be very difficult for us to return and go around by the Kasane way. Finally he relented, saying, "It's OK with me, sir. I'm velly happy to let you go through, sir. But first you must get the permission from the sergeant." He pointed us to a house on top of a nearby hill.

We drove up and found the sergeant taking his tea in his nicely appointed bungalow. It had a panoramic view over the valley and the death strip below. He was resplendent in colonial uniform, white knee socks and shorts, and starched white short-sleeved shirt with tabs on the shoulders and red sergeant stripes on the sleeve. Oh good, I thought, a white man. We'll get this problem fixed up straightaway. Uh-uh. No way, José. Tourists can't cross at Panda-ma-tenga. No exceptions are made. We could cross only at Kasane. He made us go back the way we had come.

I felt about like the time when, after four days of backpacking in from the eastern side of the Sierras, a ranger I had met while cutting across a remote corner of Kings Canyon National Park made me return the way I had come for having my dog with me. Frustration wasn't quite the word for it. We drove back through miserable shot-up Panda-ma-tenga, cursing Blackie all the way. "You got to go by

Panda-ma-tenga, boss. It's the best way, boss." Panda-ma-tenga my ass! I drove back up onto the levee and on sudden impulse turned right. "What the hell, Teddy, let's take this thing. It's going in the right direction." I couldn't face going back into the bush.

We boomed along on the graded surface of the levee top, doing about sixty. It seemed like we were flying after our morning of crawling along on the road through the forest. We were headed northwest in a straight line parallel to the Rhodesia border. If this kept up, we would come out opposite Kazungula in no time at all. We just had to hope it didn't stop somewhere in the middle of nowhere. There was a certain eerie familiarity about this structure, and this feeling grew as we drove along. "Jesus Christ, Ted. This sure does look like a freeway, doesn't it?"

"Heh-heh. It sure does. That's what I've been thinking all along."

Every once in a while we would pass a dry watercourse, where there would be a big shiny new culvert pipe under the road. "Not only a freeway. It looks like a freaking interstate under construction. But what in the hell would a freeway be doing here?"

"Man, I don't know, but right now I'm not complaining. It's going right where we want to go, isn't it?"

According to our maps, there weren't supposed to be any roads of any description at all over in this part of the country. And this thing hadn't been on any of the satellite images, either. Anything this size would have stuck out like a sore thumb on a satellite view. Besides, finding a freeway in Botswana seemed about as likely as running into a herd of wild elephants in Central Park. No, come to think of it, less likely.

We continued booming along, silently praying that our magical freeway wouldn't start petering out like so many of the other roads we had followed in Botswana. Gradually we started to come upon signs of recent construction activity. The "lanes" on the opposite side were piled with loose earth ready to be graded. A big truck appeared far ahead on the road. It was headed in the same direction as we were and it took us a long time to overtake it. As we got closer we saw that it was some kind of huge construction truck, painted Euclid green. Finally we overtook it, and as we passed it we saw a sign on the cab door. It said:

Grove Construction Co.
Paterson, New Jersey

"Holy shit!" I looked at Teddy, "Did you see what I saw?"

"Yeah. Twilight Zone. He-he-he."

We drove on for a while, passing a lot more trucks and seeing graders, scrapers, and levelers working on the other side, all labeled Grove Construction Co. Eventually we came to a big road ramping off to the right. I took it, and we soon came out into a huge sand and gravel pit that was being excavated for the road construction. We got out, stretched our legs, and refueled the vehicle from the jerry cans. A pickup was parked fifty yards down the road. A couple of guys in hardhats were standing next to it so I strolled down to say hello. They were American engineers working for Grove on this project. We ended up giving one of them a lift up to their construction camp on the Zambezi, and along the way he explained to us the mystery of the secret freeway.

According to him, Henry Kissinger had made one of his shuttle diplomacy trips to southern Africa a year or two before. His mission was to try to sort out some of the problems in this region where three civil wars were going on simultaneously and threatening to spread to the neighboring countries. One big problem caused by the Rhodesian conflict was that because Rhodesia had shut down the railroad at Vic Falls, Zambia couldn't get its copper out to market and was threatening to collapse for lack of export revenues. So Kissinger said, "no problem," and arranged for the A.I.D. to build this freeway between Francistown and Kazungula to bypass Rhodesia. A new bridge was going up across the Zambezi to make the last link. All this was being done in a very hush-hush way, because if it became public knowledge there would have been an uproar in Congress and a few other choice places as well. They had done a pretty good job keeping it on the q.t., I thought. We had been all over the country in the past couple of months and hadn't heard a word about it. It wasn't the usual American cock-up, where everybody knew what was going on except for the security guys.

The Rhodesian border post by the river had been half wrecked by a mortar attack the week before and was being manned in a temporary setup in one of the few buildings still left intact. There were a lot of people milling around in the confusion. When we finally got to the head of what you might call the line and presented our passports to the border guard, he went through the usual customs litany. "Purpose of your visit to Rhodesia?"

"Tourism."

"May I see your return airline ticket?"

"Huh? Hey Ted, do you have your airline ticket with you?"

"What? No. I left it back in camp in my kit."

"Sorry, we don't have our airline tickets with us. See, we're working in Botswana. For the U.N. We've been there a couple of months and . . ."

"Sorry, gentlemen. You can't enter without proof of return passage to your country of origin. If you don't have your airline tickets you can fulfill that requirement by demonstrating that you each have $2,000 in your possession."

"Sir, could you please step outside just a second? See, we're driving that U.N. vehicle over there," I said, pointing to our Land Rover, which had "U.N.D.P. Okavango Project" stenciled on the door. "We have to return that to Botswana, so how about letting us use that for proof of return passage?"

It worked! Far out. A border guard who knew how to bend a rule! I never would have dreamed that it would be so hard to get into Rhodesia.

On the way out of the customs compound we picked up a big black guy who was hitchhiking. The road on the Rhodesian side of the border was a great improvement. It was graded and well maintained and it ran along in a tunnel between tall trees that dappled it in shade. I asked our hitchhiker, whose massive body filled the jump seat behind us, "How far are we from the river?"

"It's 'bout a half mile or more over dat way," he said, pointing off to our left. "It curve back and forth. Sometimes it get close, sometimes far away."

"That's the Zambian border, isn't it? I hear that they shell this road from over there."

"Sometimes, da guerrillas, they shoot at de road wid de mortars. Last week dey doin' dat. Hit da border post pretty good. Hit some trucks on de road, too. Dis week pretty quiet, though. No problem dis week. Only you got to watch out for de mines."

"Mines?" said Teddy from behind the wheel.

"Yeah, da mines. Dey sneak over in da night an' put da mines in de road. Dey dig a hole and put de mine in de hole. You drive on it, boom! Got to look careful. If you see de place in de road where somebody been diggin', better go around dat place."

This gave Teddy a case of the jitters. He was imagining freshly dug places everywhere. He would jam on the brakes at every patch of shade or slight irregularity in the road, and our forward progress became a series of lurches and jerks.

After enduring this for a while, I turned to him and said, "Say, Ted, mind if I take over for a while? I'd like to get there sometime tonight."

He grinned sheepishly. "I guess I got a little nervous about the mines, huh?"

"You could say that again." We switched over and I made good time the rest of the way into Vic Falls. To hell with the mines, I wanted my gin and tonic.

So it was that a good twelve hours after starting out, we finally arrived at the Victoria Falls Hotel in our dust-coated and slightly battered Land Rover, just as the last of our spare tires went flat in the parking lot. We grabbed our rucksacks and on legs rubbery from all those hours of bouncing around on the road, walked up to the entrance of the hotel. It was a huge victorian colonial pile, a monument in testimony to Empire. At the entrance we were greeted by a black giant of a majordomo—or, to be more precise, we were stopped. He was shorter than us but as wide as he was tall. He wore what looked like the uniform of a field marshal in the Freedonian Army: a peaked hat with visor encrusted with gold braid, a double-breasted jacket draped with epaulets and decorated with a row of medals across his ample chest, and trousers with red satin stripes down the seams. We, on the other hand, were decked out in our usual field finery: boots, shorts, scruffy shirts, and floppy brimmed hats, and we were caked with many layers of dust streaked by rivulets of sweat. In, short, we weren't looking our best.

"Ahem. Will you gentlemen be looking for the bar?"

"No, we want a room. We plan to stay here."

He gave us a final look-over and, reaching some inscrutable decision, said, "Follow me, gentlemen." He personally ushered us all the way across the ornate lobby to the reception desk, where he stood impassively behind us until he had made sure that we had completed all the formalities and received our key. We looked around the elegant old lobby, feeling dazzled and out of place. It was a beautiful set of public rooms in the old style, lined in intricately carved dark woods. One area had been set aside with tables at which a group of well-dressed ladies were having their high tea, being helped to creams and cakes by a small platoon of waiters plying dessert caddies of glass framed in gleaming brass. In another alcove lined with leatherbound books and with tables strewn with magazines a few men in tweeds sat in highbacked leather chairs reading newspapers. To our chagrin, we discovered that we had been leaving a trail of dusty footprints on the Oriental carpets.

Our room was equally classy, making us feel shabbier than we already did. We didn't feel comfortable sitting down on any of the furniture. Ted went into the bathroom to take a shower while I called room service for gins and tonics. When they arrived they were just like the one in my dream. I ogled them for a while before picking one up and slurping it down in not so many very delicate sips. I called back and ordered another before going for my bath. Teddy was dancing around the spacious tiled bathroom, looking pink and chirpy and very clean. "Marvelous, simply marvelous," he said. "I haven't felt this clean in months. Even my hair squeaks," he said, demonstrating.

I looked in the bathtub. "Maybe so, but it looks like the Red River Delta in here."

After bathing and changing into clean clothes we felt greatly revived, so we went down to the bar for a drink before dinner. The atmosphere there was a bit more congenial for us, there being a comfortable mix of riff-raff among the gentlemen. For dinner we chose the tourist braii set up on the back terrace rather than suffer the dignities of eating in the formal dining room.

The hotel is situated on the edge of the gorge opposite the falls, and the view from the terrace was spectacular. At Victoria Falls the Zambezi flows off the side of the gorge rather than its end, as waterfalls usually do, which is what makes these falls so special. They are made up of a series of cascades sheeting off the sheer edge of the cliff for more than a mile along the length of the gorge. We were there at the right time that night to see the sun setting obliquely through the mists, producing a host of spectacular rainbows and other optical effects.

The braii had a good selection of meats, both domestic and game, as well as a huge selection of salads and curries, which we particularly made pigs of ourselves over. It was followed by an exhibition of native dancing. After taking my seat for the performance, I let out a satisfied belch, and leaning back, promptly fell asleep.

The next morning we took a tour of the falls. Near the end of the gorge there is a region filled with a thick jungle of tree ferns, vines, and wildly colorful tropical plants, which flourish in an ecosystem created by the mists from the falls. None of those plants was to be seen anywhere else in that whole area. Such isolated habitats have always fascinated me. Whenever given the slightest toehold, life of some sort seems to thrive and fill all possible space. Witness the clam colonies that live at abyssal depths around the "black smoker" vents along the East Pacific Rise. I always ask myself the question: How did they get there?

We went into town and had our Wimpys, which, sadly, were not anywhere close to the ones in my dreams, and then we did some shopping. The shopping district was the image of a High Street in a provincial English town. The shops were abundantly stocked with a range of goods wider than any we had seen since leaving New York. This was all rather surprising considering that at the time there was supposed to be a total British trade embargo on Rhodesia. We stocked up on meat and vegetables, and at one store picked up a couple of bottles of scotch. They were labeled with some bogus brand name like "Glen Hogan" and inscribed "Product of Rhodesia." I opened one and took a sniff. "Hmm," I remarked, "either the embargo has a few leaks in it or they have recently discovered peat beds in Rhodesia."

The return trip to Ngwezumba via Kasane was a mere three and a half hours of relatively easy driving. The crew had arrived back at camp just before us and looked equally refreshed. That evening Deacon prepared for our dinner potatoes and the mutton chops we had bought in the proper British butchery in Vic Falls. It was delicious. The mutton chops were a heavenly change. We were sick of eating filet mignon day in and day out. When Deacon came to clean up, Teddy said, with his bogus Colonel Blimp shtick, "Splendid meal, Deacon. Simply tip-top. My compliments to the chef." Deacon beamed.

I was feeling equally expansive. "Next time, though, try sprinkling them with a little lemon and herbs de Provence."

"What you say, boss?"

"Never mind." I was already thinking about how the fresh eggs and bangers we'd bought would taste for breakfast the next morning. It's amazing what kind of food fetishes certain types of deprivation can bring on.

Our ultimate crisis with the elephants occurred that night. I was sitting at our mess table reading *Oliver Twist* by the light of a Coleman lantern. I had just got to the part where Bill Sikes was trying to drown his dog, when a tremendous uproar broke out just outside of camp. There was a terrific crashing followed by the bellows of the trumpeting of more than one elephant. I jumped up in alarm and Teddy came running out of the tent. I could see, just at the edge of the lantern light, trees shaking in the dark as though in the grip of some monstrous force. Suddenly the rear end of an elephant appeared in our clearing and spun around, followed by another of the gigantic beasts. Lit starkly by the light of the lanterns, the two bull elephants bellowed and, trunks flailing, reared at each other, butting and kicking and stabbing with their huge tusks. Wheeling around, they continued fighting right into the center of camp. Teddy and I both instinctively stepped back, but there was no place to go except out in the black unknown of the forest where the lions were prowling. I felt like an ant in a sumo ring. The huge behemoths were intent only on doing in each other, but what was to prevent our getting stepped on or otherwise incidentally crushed in the course of their mortal combat? The electric lights went out when they ripped out the wires, but Teddy grabbed the nearest lantern before it went over so we weren't plunged into complete darkness. He ran around holding it up in front of himself like an amulet to ward off the elephants. Our participation in the elephant fight was probably all over in thirty seconds or less, the time it took them to come in one side of the camp, wheel about a few times, parrying and thrusting, and go out the other, leaving a swath of destruction to record their passage. But for us, that had been an epic thirty seconds. Gradually the sound of their fighting receded into the forest. Realizing it was finally all over, I started breathing again.

Teddy and I walked around with lanterns to survey the damage. We had been very fortunate. The elephants had come through the middle of camp, between our camp and the crew's. Deacon's cooking area was pretty well trashed. Our bathtub was mangled, and the little gasoline powered generator that we used to run the electric lights and charge batteries had been kicked about ten feet. The cases of several of the batteries had been cracked and were leaking acid. Teddy righted the generator and cranked it. It started up with a roar. Our tent was hanging at half mast because some of the stakes had been pulled up, and the guy wires for our radio antenna had been ripped loose. Luckily, we had not parked any of the vehicles in the center of camp that night, so they had been spared.

We hadn't been paying any attention to the crew this whole time. They had stayed clustered around their fire, which was stoked up into a raging inferno. Eventually they came creeping over to look at the mess. I took Blackie aside. "It's time for us to be leaving this place, Blackie. It will take two days to retrieve the instruments and get everything packed up. I suggest that we leave on Friday. That will get us into Maun for Independence Day."

He nodded gravely. "That very good plan, boss. Very good. Very wise plan." He went back to pass the news on to the rest of the men.

The next morning when I tuned in the radio-telephone at the usually scheduled time for Geological Survey communications, I received a message from headquarters in Lobatse. They had received a telegram for me from New York, and proceeded to read it out to me. Our reception was so bad that I couldn't make out half of it, so I switched on the microphone and said: "Scholz here. Could you read that message again please. Our reception is very bad. Elephants came through camp last night and knocked down our radio mast. Over."

There was a moment of silence. Then an incredible uproar came over the radio. Everybody in the country seemed to be talking at once, jabbering in a mixture of English and Afrikaans. Gradually it dawned on me what was going on. The Afrikaner well drillers. The bloody Afrikaner well drillers! Those bastards had been running a sweepstakes on us the whole time we had been in the Chobe, betting on which day the tinhorn Yankees would cut and run! They had been listening in on the radio-telephone each morning, betting charts at the ready, just waiting for us to announce that we were bailing out. I wondered how they would sort out the winnings now.

13

We set up a fly camp at our usual spot outside of Maun. There were a lot more camps dotted about in the scrub than usual, many groups having come into Maun from the outlying districts for the Independence Day celebrations. Riley's was jam-packed on the eve of the big day. Across the street in the big square a soccer field had been freshly chalked, and a sort of arena ringed with bleachers had been set up under a big olive-drab canvas tent. Crowds of people were strolling around the square and wandering in and out of Riley's, creating an atmosphere of building excitement. I was reminded of rural county fairs my parents had taken me to as a boy. All that was missing was the Ferris wheel and the other rides, and the carneys selling cotton candy and hot dogs.

We went into town early the next day and established ourselves at a table in Riley's garden before the main crowds arrived and took up all the available sitting space. After a while the soccer match started up across the street and Teddy and I went over to watch. It was a schoolboy game, being played vigorously on the dusty field by teams decked out in somewhat rag-tag but still respectable uniforms and presided over by referees in striped shirts. The only thing nonregulation about it was that all the participants were barefoot. This did not seem to handicap their play in the least; they were scrambling around with great energy and verve. A sizable crowd assembled around the field, shouting "Pula! Pula!" whenever a goal was scored or some other exceptional play made. Pula literally means "water" in Setswana, but it is also the motto of the country and is used universally as an expression of greetings, joy, or other general encomiums. So Ted and I, joining into the spirit of things, started yelling "Pula! Pula!" too.

After the game we repaired to our table and had a few beers while we watched the people parading along the street. Under the trees by the river a group of people were starting to roast an ox over a huge barbecue pit. They had rigged it up on a spit that was being cranked on each end by sturdy women. Nearby, huge tubs of mealies and pumpkin were being prepared. Women were carrying loads of supplies over to the cooking area. As in most of Africa, all the heavy loads in Botswana seem to be carried by women on their heads. They carried great bundles of poles of firewood twice as long as their bodies and as wide as their waists in graceful insouciance, the ends of the poles barely swaying as they strode along. With just a little pad the size of a folded up bandana for protection, they would stroll along with an open bucket holding four or five gallons of water perched on their heads without spilling a drop, one hand raised to touch the pail for balance. They accomplished this with a sinuous, hip-swiveling stride that kept their heads moving forward in an absolutely level trajectory. I've read somewhere that this method of carrying requires less expenditure of energy per unit load than any other, being far more efficient, say, than carrying a load in a backpack. We saw young girls of seven or eight walking around proudly practicing with a coke can or packet of cigarettes perched on their heads. Teddy was greatly amused to see this, saying, "Look ma, no hands. He-he-he."

Groups of ladies strolled back and forth arm in arm, chatting. They were resplendent in fine dresses that appeared to have been made specially for this occasion. These were long, full robes with puffed sleeves, and they had elegantly wrapped turbans to match. Their dresses all looked as if they had been cut from the same pattern, but each was made distinct with cloth printed with a different swirling design of bright colors that contrasted finely with their shining black faces. They were batiks from India and Sri Lanka, sold in the Indian shops and in outlying areas by traveling merchants.

An uproar of cheering and shouting suddenly broke out, and crooking our necks to look over the shoulders of the crowd that had gathered by the garden wall, we saw a long line of children running toward us, screaming and shaking toy spears decorated with streamers. At the head of this mob ran a most peculiar figure: a wizened old man, his body painted in boldly colored patterns, wearing a short spiky grass skirt and a grotesquely carved mask with a feathered headdress. He was waving a big feather-bedecked rattle in one hand and an assagai in the other. "Holy shit!" Ted shouted excitedly. "A witch doctor! A real witch doctor." And with that he ran out of the garden and chased after the cheering mob.

I followed at a respectable speed and distance. The witch doctor had disappeared into the tented structure, around which a large crowd had gathered and was milling about trying to get in. Teddy was nowhere to be seen. I got into the back of the crowd, and as the last people climbed the steps and filed in, I wedged myself in among the standees in the last row. Rows of bleachers descended from all sides of the tent down to an elevated platform like a boxing ring in the center. The arena was absolutely packed, rows upon rows of black faces gazing and cheering at the witch doctor on the stage who was dancing and gesturing with his rattle and spear. I was very nervous about being in there, feeling like an interloper at somebody else's sacred ritual. I had no idea how seriously the Tsetswana still considered their old religion, but one thing was for sure: This performance was not being enacted for the benefit of white tourists like me. The only reason I was there at all, timidly peeking around, was that I felt sure Teddy was in there someplace.

It didn't take me long to spot him. Some flashes lit up the darkened tent, and peering between the shoulders of the crowd in front of me, I saw Teddy. There he was up against the stage in the front row, grinning his head off and blazing away with his Instamatic! Aside from mine, his was the only white face in the place. The witch doctor, a real vaudevillian, was taking it all in good humor and was actually making Teddy part of his act, jumping and shaking his rattle at him to the roars of laughter of the crowd. I had a feeling that all this good humor wouldn't last too long, though. Sooner or later the witch doctor and the crowd were going to get annoyed at Teddy horning in on their act. I had to get him out of there before trouble started.

I found my way out of the tent and ran around to the side nearest where Teddy was. Opening a flap, I edged my way into the crowd and squirmed through until I was next to him. I grabbed him by the arm and said in his ear. "Come on, Ted. Let's get out of here."

"What do you mean? This is fun. I don't want to go yet."

"Bullshit. We're leaving now." Grabbing him forcibly by the armpit, I pulled him out through the crowd.

On the way back to Riley's he complained, "What the hell was that all about, Scholz? We were having fun. The witch doctor is a great guy."

I was a bit shaken and pissed off at Teddy about the whole affair. "Damn it all, Teddy, you can't be barging in on things like that. That show wasn't for tourists, you know. You're damn lucky you still have your camera, to say the least."

Ted was pretty sore with me about that. Years later, after many other trips to remote places, I realized that he was right in a way; he probably wouldn't have gotten into any serious trouble. No matter where he is, he has an amazing facility for getting on with the local people and making friends with them even when he hasn't the slightest knowledge of their language and customs. Once in Dushanbe he disappeared for most of a day, finally turning up in the evening with a group of Tadjik herdsmen who had become his buddies after spending the afternoon together drinking tea in one of the tea pavilions by the market. From them he had somehow learned not only where to get the best meat and vegetables, but where to obtain contraband electronic parts on the black market, even though he knew but half a dozen words of Tadjik, most of them obscene. In situations like the witch doctor show I tended to get uptight, and my discomfort could be sensed by everyone, so my presence became quickly resented. With Teddy, on the other hand, his friendly and outgoing disposition was equally apparent to all, so he could commit the most egregious faux pas without generating the slightest enmity. This talent of his was one of his greatest assets in doing fieldwork.

By early afternoon the garden at Riley's had filled up with noisy celebrants. Almost everyone we had ever met around Maun was there. Joe, accompanied by Clyde on this occasion, and Dave and Sarah Potten joined us at our table on the verandah, where we had moved to get out of the hot sun. I even saw the dour Swedish hydrologist, who didn't usually go in for socializing of any sort, drinking a beer in the garden with a small group of what looked to be his countrymen. Most of our crew were enjoying themselves in the bar. They had been salting away part of their subsistence allowance for the past month for this big day on the town. There were a lot of boisterous greetings among people who had come into Maun from remote settlements like Ghanzi and were meeting old mates they hadn't seen in many months, often since the last Independence Day celebration. Everyone was being very friendly to old pals and strangers alike, so it was becoming one big party.

A group of the elegantly clad ladies had entered the garden and were beckoning to me. I gave them a quizzical look and one shyly detached herself and came over to our table. She pointed at my camera, then back to the group of ladies, who smiled invitingly. Ah, they wanted me to take their picture. I went over and lined them up in two rows below the big tree in the corner of the garden. They posed, primly but proudly in their brightly colored gowns, while I took several photographs. Suddenly I was grabbed from the rear by someone screaming, "No apartheid here! No apartheid here!" I spun free and confronted a short, wild-eyed young black man sporting wide horn-rimmed glasses and an afro. He kept shouting his "No apartheid here!" mantra and tried to grab my camera out of my hands. I wrested it back, and mindful of the scene that was being created, told him to buzz off and made my way back to our table on the verandah. He followed me, screaming various anticolonialist imprecations the whole time. He managed once again to grab my camera strap, and we were having a tug-of-war over it when Blackie and Boy emerged from the bar, where they had been fetched by Samuel. Picking my tormenter up by the elbows, they detached him from my camera strap and frog-marched him around the back of the building. He was not seen again for the rest of the day.

"What the hell was that all about?" said Teddy.

"You got me. He must be a recent graduate of Patrice Lumumba University, or something."

"The what?"

"It's in Moscow. That's where they teach African students how to be good little revolutionaries."

Later I went over and got the mailing address of one of the ladies. The pictures turned out very well. A few months later, when I mailed them some prints, I imagined the delight with which they would be received. Photographs were not everyday items for those people; it would be unusual for a family to own one.

A big crowd queued up at the barbecue pits when they took the ox off the spit and started hacking it up. We joined the line and collected our platefuls of bloody hunks of meat and big scoops of mealies and pumpkin. It was a primitive banquet, but after watching the cooking for so long everyone was ravenous by the time it was finally served. We wolfed it down eagerly with a minimum of decorum, the grease dripping off our chins and elbows, and washed it all down with a few beers.

In our postprandial stupors we dozed off in our chairs awhile after the meal. We were startled awake by a wiry beanpole of an Afrikaner who came up to our table, exclaiming, "Say, aren't you the Yanks that were doin' the seismic survey up in the Chobe?"

When we acknowledged this, he said, "Well I want to shake your hands. Piet Onderdonk, glad to meet cha. I'm a well driller in the Ghanzi farms. A good job of it, that was. Ja, you got the joke on us, too. When you reported in about the elephants being in your camp, just as matter of fact as if they'd been stray cows, I got the biggest laugh I'd had all year." He proceeded to fill Joe and Clyde in on our adventures, as pieced together from our radio reports, and the parallel story of the well driller's lottery.

"Ja, ja, we'd been following these boy's progress for months. Not much to do out in camp, see, so the radio-time is a big entertainment for us. There's the official one in the morning, o'course. Then we drillers have our own unofficial one at night when we jaw away at one another. We had a good laugh when these blokes reported in that they were movin' up to Ngwezumba. Nobody gave 'em the slightest chance of sticking it, see. We were all speculatin' about how long they'd last, when Junius Bronk came up with the idea for a lottery. Ten-rand stakes, everybody picks a number out of a hat. I got day five, so I thought I was sittin' pretty. So who gets the pot? Old Titus from Mahelapye. He drew day twenty-five, the last day, so nobody figured he had a prayer, and he was moaning and groaning about it. But these boys beat him by a good five days." He slapped his leg, laughing uproariously. "It was lovely. Bluidy lovely."

"Ja," he said, turning to me, "you did a good job on old Ranger Dick, too. We heard he got a reprimand from the ministry over it. He's a right bastard, he is. I loved to see him get his comeuppance."

We bought Piet a beer and told him a few more tales of our Chobe adventures. When he got up to leave, he said, "Well, I sure am glad I ran into you boys. We drillers are having our end of season braii in Frauncestown on the fifteenth. Try to make it if you can. Everybody will want to meet you and stand you a beer. Hell, we'll be talkin' about this one for years."

The darts tournament started up as soon as the crowd in the bar thinned out enough to give us space to play. Ted and I had entered, and we spotted our initials up on the big draw board. The first two rounds went slowly, because there was a big field entered. This was mainly for warming up, while the weak players were being eliminated. Teddy got knocked out in the early part of the middle rounds and told me he felt like packing it in for the night. "OK," I said, tossing him the keys, "I'll get a lift back to camp from Joe." I had reached that intermediate state of inebriation where my darts playing was at its best, sober enough to be able to concentrate on the board and drunk enough so that my delivery was smooth and sure. It was as if I could see little dotted trajectories going from my hand right into the doubles ring. I managed to maintain that condition for quite a while, until the final rounds, when my fatigue and blood-alcohol concentration conspired to collapse my level of play.

It was nearly closing time before I was finally knocked out of the tournament and wandered out to the verandah, looking for Joe. He was sitting at a table with a couple of his pals, telling a long-winded but funny yarn about the efficacies of leopard fat as an aphrodisiac. "Ah, there you are, me boyo. You were jolly good tonight. You were looking right unbeatable for a while there. Come on, let's go back to my camp and have a nightcap. I've got a couple of girls to go with us."

We went out to the nearly empty parking lot and climbed into his pickup. Two girls slipped out of the darkness and climbed in with us, one wedged in between us and the other perched on my lap. Although my memory of that evening is a bit dim, I don't recall having arranged anything with Joe about taking some girls back to his camp. Still, the feeling of this plump girl wiggling on my lap was beginning to revive me, and I was starting to think that maybe it wasn't such a bad idea after all.

Joe drove out of the parking lot, just remembering to switch on the lights when we reached the main road. I heard a commotion behind us, and looking back, saw a truck, lights blazing, roaring out of the parking lot after us. Just as we pulled onto the main road it drew abreast and I saw that it was the police paddy wagon, with cops hanging all over it shouting and waving at us. We roared off down the road together and then the paddy wagon swerved toward us, as though to cut us off, but instead crashed into our right front fender. Together we came to a metal-rending halt. The drunken cops piled off and swarmed around us. They got our doors open and reached in, grabbing the girls and trying to pull them forcibly out of the cab. The girls resisted, screaming at them. Joe yelled, "What the bloody hell is this! Who the hell is in charge here?" This slowed the cops down enough that we managed to get the doors closed again. Joe looked at me and said, "Stay here and keep quiet. I'll handle this. I know how to deal with these boogers."

He climbed out and went around to inspect the damage. The whole right front of his pickup was stove in, and the paddy wagon had not fared much better. "Look vaat you've done to me bloody truck," he said, in much aggrieved tones. "Who's responsible for this mess, hey?"

Meanwhile, the driver of the paddy wagon had got out and was inspecting the damage himself. He put his hands to his face in dismay, and then yelled at the other cops, who were still milling around my door jabbering at the girls. They all suddenly became much quieter, sobering up quickly at the realization that their drunken idea of rescuing the girls from our clutches had backfired and landed them in some hot water. They had smashed in someone's truck, and worse, the police department's prized paddy wagon. How were they going to explain that away?

The paddy wagon driver, who seemed to be the man in charge, had a bright idea. "It is your fault, sir. You were driving without headlamps on. This is a violation. I will have to give you a summons."

"Don't give me headlamps," said Joe. "I turned 'em on when I got to the road. And that's no good reason for driving into me and wrecking me truck. Just look vaat you've done, man."

This wrangling went on interminably, and was leading absolutely nowhere. It didn't look like we would ever be getting back to camp at this rate. Feeling bored and disgusted with the situation, I climbed out of the cab and had a walk around to calm my frustration at being cooped up here at this ridiculous accident scene instead of being snug in my bed in camp where I belonged. When I returned to the pickup I found that the girls had disappeared, which did not in the least disturb me, since any interest I had had in them had long since flagged.

Joe was still arguing with the paddy wagon driver when another figure arrived, walking out of the darkness carrying a lantern. When he came under the streetlight, I saw that he was wearing the uniform jacket of a sergeant of police. The effect was somewhat marred because it was buttoned askew and worn over what appeared to be his pajamas. He was also walking with a decidedly uneven gait.

"Ah, sergeant," Joe exclaimed, "Good you could come, sir, good you could come. Now we can get this little matter straightened out. You see, we've had a bit of a collision here, between the police truck and mine," he said, leading the sergeant over to the trucks, which had by then been detached from one another. Before the paddy wagon driver could intervene with his story, Joe continued, "Vaat we've got to have here, sir, is a proper report. A proper accident report. Have you got your notebook, sir?" The sergeant hadn't, but the paddy wagon driver produced a notebook and the stub of a pencil, which the sergeant accepted and opened up to a blank page.

"Ja, ja, good. Vell, vaat ve need is a proper diagram of the scene of the accident. If you don't mind, sir, to have your men pace off the area so that ve can have a proper diagram." So the sergeant dispatched his men to pace off the distance to various trees, posts, and other objects in the vicinity. Trying to concentrate on making long even paces only accentuated the staggering of the policemen, so they proceeded to their assigned landmarks in carefully paced out weaving paths. When they returned to report their findings to the sergeant, Joe exclaimed, "Ve got to check these figures, sir. Accuracy, that's the thing. The report must be accurate." So another set of policemen was set out to check on the first, and of course they returned with completely different answers. This charade of police work was carried out several more times, the results becoming increasingly less conclusive. Then Joe got on to the sergeant about the need to measure angles to things. The concept of angles was completely outside the realm of experience of the sergeant, so Joe began a tutorial on the subject by drawing diagrams in the dirt beneath the streetlight, while the policemen gathered around, wagging their heads in bemusement.

I was at first greatly amused by Joe's masterful handling of the situation, though I was careful not to let any hint of my hilarity show. All this activity that he was orchestrating was a brilliant diversionary tactic. The policemen were being kept busy doing various tasks, and had long since forgotten any grievances they might have had with us, even their original motivations in the incident. The big flaw in this strategy was that it had no logical conclusion, as far as I could see. As it went on and on and on, I became more and more disgruntled. I was getting thoroughly fed up with the whole situation.

All this senseless activity ceased with the arrival of yet another official on the scene. This was a skinny little man rigged out in a safari suit and a topee that looked several sizes too large. With his tiny wire-rimmed spectacles he looked like Mahatma Gandhi dressed for tiffin. He strutted up pompously, waving a swagger stick at the sergeant and demanding his report. The sergeant came to stiff attention and presented his tattered notebook. The inspector, as I took him to be, glanced at the sergeant's scribblings and demanded a new investigation. Soon the police were sent off once again on their pacing missions.

This was too much for me. "Bullshit!" I raged, and started off to give everyone a piece of my mind. Almost immediately a pair of enormous arms circled me from the rear and, pinioning my arms to my sides, lifted me off the ground. A deep voice from behind rumbled, "Yoo maak da disrespek. Yoo unner arrest." Stunned, I was bodily carried by the huge policeman over to the paddy wagon, where he tossed me into the cage in the rear and locked me in. I was so enraged that I literally saw red. Jumping around the cage like a monkey, I screamed and bellowed to be let out. Just then, Teddy showed up. He had been roused out of bed by Blackie, who had witnessed the beginnings of our altercation with the cops and had hurried back to camp to fetch him.

"Cool it, Scholz. Simmer down."

"Jesus, Teddy, these idiots have arrested me! This is an outrage. I'll be damned if I'm going to be thrown into some pisspot jail. Call the U.N.D.P., call the freakin' Ambassador! Get me out of here!"

"Take it easy, will you? Just calm down. Let Joe handle things. He knows what he's doing. Screaming bloody murder won't help anything."

I quieted down but I didn't calm down. My rage transformed into a deep sulk, as I crouched in the corner of the cage and listened to Joe's interminable negotiations with the inspector for my release. Finally, after what seemed an eternity, they came around to the paddy wagon and let me out.

"Go on back to camp and get some sleep," said Joe. "We'll settle this in court in the morning. Don't worry, everything will be all right."

Joe picked me up the next morning and we drove into Maun for my court appearance. When we pulled up in front of the whitewashed rondavel that was the D.C.' s courthouse, Joe pulled a package out from behind the seat. "I brought a bottle of whiskey for the D.C. Don't worry, he won't cause any real trouble. But a word of warning to you. Vaatever you do, once ve get in there, don't laugh."

The inside of the building was fitted out just like a miniature English courtroom, complete with gated wooden railings and benches, all neatly tucked into the circular floor plan. I could imagine some British firm somewhere in the Midlands that specialized in courtroom fixtures for tiny courts in remote colonial outposts. This was the rondavel model. A door in the rear opened and the D.C. entered. Good lord, it was the "inspector" of the night before! He came up through the gates and said a few words to Joe, then turning to me, he gravely stated, "You have been accused of a serious crime against the government. Please enter the court and you may argue your case."

He led me up to the bench, where he installed himself. In elaborate phrases he proceeded to swear me in and recite the government's case against me. The reason for Joe's warning became instantly apparent. His pompous gestures and pontifical style of speech were a hilarious caricature of British dramatic courtroom scenes. He must have rehearsed them from listening to BBC detective programs late at night and then improving on their impressive effect by adding his own flowery embellishments. It was a major effort to listen to him with a straight face. But that, I realized, I must do without fail.

I requested that he explain to me the exact statute under which I had been accused. He passed over the book of statutes and pointed out the appropriate paragraph. I was accused under a law that made it a criminal offense to make any public utterance that smacked of racism, colonialism, or disrespect for the authorities. In, short, it was a blanket law that could be used to cover any sort of perceived insult, all of which was completely up to the interpretation of the accuser. Reading further, I found the chilling statement "Conviction could result in a sentence of up to a year in prison." There was no way out of this but by abject apology. Adopting what I hoped sounded like a very sincere tone of voice, I explained that in my country the offending term was often used in everyday conversation to express surprise or lack of understanding. I profusely apologized if my use of it had been misunderstood as a lack of respect, but that I most certainly hadn't intended it that way. I simply had not known the connotation that this term had in this country.

He solemnly accepted my apology, admonishing me against any similar outbursts in the future. With a deep sigh of relief, I rejoined Joe and we left the courtroom, his package remaining behind on a desk. "Ja, ja, everything's OK, see. But who's going to pay for repairing me bloody fender?"

14

After recuperating from our Independence Day celebrations we headed up north to establish our last camp, in the Moremi Wildlife Reserve deep within the Okavango. The Moremi is a triangular peninsula of dry land that extends far into the swamps at the northern end of the Okavango. It is entirely within Batawana Tribal Territory, and is operated as a commercial enterprise catering to air safari groups that fly into the Khwai River Lodge and from thence are conducted through the reserve in guided parties. Those who take such tours will not be disappointed: The reserve shelters a wildlife population of great density and diversity, with the riverine and savanna populations in close juxtaposition. The track from Khwai follows along the edge of the dry land bordering the swamps. It runs through a sandy forest that is much more thickly forested and with a greater variety of trees than found in the dry veld of the Chobe. In many places it is verged by swamplands that are bright green in lush foliage that almost engulfs the many narrow watercourses and pools. Giraffes and elephants can be seen tranquilly grazing in the forest, and the swamps abound in waterfowl of so many varieties that it is breathtaking at first sight. A short while after entering the reserve on the road from Khwai, the road passes the Dombo Lagoon, a large hippo pool surrounded by grassy meadows grazed by large mixed antelope herds, including the beautiful red lechwe, a rare antelope whose only habitat in Botswana is the Okavango. This kind of sight inspires thoughts of an Eden in the primitive cast imagined in the jungle paintings of the douanier Henri Rousseau.

Camplife there in the Moremi was a welcome contrast to our precarious existence at Ngwezumba. We set up our camp in a sandy forest on the banks of a clear stream that ran along in its sandy bed through thick growths of papyrus and other rushes. Here was good fresh drinking water; no longer would we have to make 150-mile round-trip treks for water that tasted of drum. Compared to what we had been used to over the past few months, it was refreshingly cool in the shade of the forest. We also soon found that although the Moremi was well populated with elephants and other wildlife, they were distributed throughout the forest, where there was a plenitude of water, and so the nights were quiet and peaceful, free of the nocturnal mayhem that we had been subjected to each evening at Ngwezumba.

Camp hadn't been set up for more than half an hour when Teddy grabbed his towel and went down to the river for a dip. A minute or two later he came running naked back up the trail from the river. "Jesus Christ, something bit me!"

"What was it? Was it a snake?"

"No, it was a fish. Just a little fish. Came right up and took a nip out of me. Look." He turned around to show me, and sure enough, there was a little bite taken out of his right buttock. He was bleeding pretty freely, so I got out the first-aid box and patched him up.

"It must have been a tigerfish. You're lucky it was a little one. If there are tigerfish around we can't get in the water. They look like barracuda and have a set of teeth like piranha, only bigger. I didn't realize they came down this far to the edge of the swamps."

"No shit, they got tigerfish here? Don't worry, I'm taking showers from now on. They got crocs here, too?"

"I don't think there are any crocs left. Poachers got 'em all in the '60s. Thomas was telling me about those days. He spent a few years in the croc poacher trade. I've heard there are plenty of green mambas, though. They are swimmers, and very nasty. Somebody told me that the original Riley bought it from one of those back in the '50s. You got to watch out when you're going through the reeds. That's where they hang out, or so I've been told."

For the final stage of our project we were going to set up our seismic network here in the northern Okavango, to fill in our coverage of the area between our earlier network in the Chobe, to the northeast, and the one we had operated around Maun, to the south of us. It was of some importance to run stations in the central part of the Okavango to get some data well to the west of the border faults that our data were now confirming were the main seismically active features in the region. I was also hoping to pick up some seismicity that might be occurring on the faults that bound the far side of the delta. The Moremi gave us the best access by land into the Okavango, but at least half our stations would have to be installed by water. We had arranged with the U.N.D.P. to supply us with a boat and guide to take us farther into the swamps. That evening, after a four-day journey from Maun, the boatman arrived at our camp, pulling up to the bank of the river in an outboard-powered skiff.

The next morning we loaded up the boat and headed upstream to put in our first station. The boat was a simple 15-foot aluminum-hulled craft equipped with a 25hp outboard. With me, the boatman, and Teddy on board there was only enough space left for the equipment for one seismic station, a rucksack with lunches, and a small supply of emergency gear. Simon, the boatman, was a very taciturn sort. He would speak only in response to direct questions, and then only in a few monosyllables. After a short while we gave up trying to engage him in conversation.

The river alongside our camp was one of the few streams navigable by a boat of even this small size. As we followed it deeper into the delta, it wound around through banks walled in by thick growths of papyrus twelve to fifteen feet high. The stream was not wider than ten or fifteen feet in most places, so there were few places wide enough to turn the little boat around, and we often couldn't see anything out of the narrow green tunnel we were traveling in except a strip of sky and the tops of occasional overhanging trees. We passed myriads of tributary channels, but the boatman unerringly steered us along the main course, and we never found ourselves in impasses. I was very impressed by this feat of navigation, since I myself could not recognize a single landmark among all the channels meandering through this sea of reeds. They all looked the same to me. By the time we had gone about five miles I had given up trying to figure out where we were on our maps. With our limited view of the sky I was even in doubt about what direction we were going in at any particular time. Before starting out, I had explained to our guide where we wanted to go, and in his terse way he had indicated that he understood. We were going to have to rely on him, though, to tell us where we were once we reached our destination.

Every few miles the stream would open out into a lagoon with broad lily pad–fringed pools lying within low regions of marsh grass. These lagoons were formed where old river meanders had

been cut off by a new channel, leaving open marshy spaces in between. Often they were bordered on several sides by low sandy islands covered with open forests of acacia and mopani trees. Along
the fringe of the land were dense brushy thickets that were teeming
with birds. These were like avian apartment houses, thick with ibis,
egrets, pelicans, storks, and hornbills of various types, all roosting
in their separate sectors, so that one often saw more bird than bush.
Kingfishers dove from the higher trees, snaring small fish in the
shallows, while from the highest branches fish eagles surveyed the
scene below. All this birdlife was testimony to the abundance of
fish, which we could see swimming in schools in the clear water.

There was a different population of wildlife to be seen within the
narrow, papyrus-walled channels. Swallows and other insectivo-
rous birds zipped along over the river surface which teemed with
an incredible variety of their prey. Turning sharp corners along the
many bends of the stream, the skimming birds looked as if they
were scooting through an aqueous bobsled course among the
rushes. As we putt-putted along we would often pass green mambas and other snakes swimming in the water, which we watched
going by with horrid fascination. This wasn't the place to be languidly trailing your hands in the water. There were many turtles
and salamanders of various types and colors, and once Simon
pointed out a small crocodile, no more than three feet long, sunning
itself on a mat of reeds. As we went by it slithered off into the water.
So the crocodiles hadn't been completely annihilated by poaching,
after all. Perhaps now they would make a recovery.

In one of these narrow spots we came around an abrupt bend and
saw, just ahead of us, a group of hippos cavorting in the stream. At
that point the stream was only about ten feet wide and the six of
them were blocking the entire channel. Noticing us, they turned
and goggled and started madly wiggling their little ears. The big
bull facing us opened its enormous mouth wide to threaten us with
a display of his huge stumpy tusks. I suddenly became acutely
aware of the flimsiness of our little boat, which looked as if it would
comfortably fit into the gaping mouth of the lead bull hippo. There
was no space for us to turn around in the narrow channel. What to
do? As I was flitting through various implausible options in my
mind, Simon shouted, "Hold on," and, revving the little outboard
engine flat out, steered directly toward them. Teddy and I gripped

the gunwales as the prow lifted and we sped toward them and what looked to be an inevitable suicidal collision. Going full speed, we were quickly upon them. Just at the last moment, with amazing swiftness and agility, they dove. As we roared over, I could see their immense bodies go by just below the boat's hull, with them peering up at us through a cloud of bubbles. Simon chuckled. We went limp. What an idea for a theme park ride! Hippo Chicken. My only second thought was: What if the water hadn't been quite deep enough for us to clear the hippos?

We installed two of our stations on sand islands, going as far into the delta as was feasible for a six-hour return boat trip, about as long a trip as we wanted to try, since we were in no mood to risk a night out in the swamps. The stations were at low-magnification sand sites, but we had no choice in this area. There is no rock to be found anywhere within the delta. If my idea of the structure of the delta was correct, that it was a down-dropped sand-filled rift valley, then the thickness of the sand beds in the delta was probably several times that found elsewhere in the Kalahari. I was no longer so acutely concerned about this problem, anyway. Our network in the Chobe had already allowed us to collect enough data to ensure that the project would be a reasonable success. Anything we picked up on this side of the faults would be gravy. I was already pretty sure that when I had finished analyzing the data back in the lab we would be able to prove the main points of my initial hypothesis.

The remainder of the network could be serviced by Land Rover, by far my more accustomed and preferred mode of transport. There were two well-established tracks in the Moremi, one along the northern edge of the "peninsula" and one along the southern, so we had no trouble in spacing out our remaining stations to form a network with a good recording geometry. Although there were plenty of elephants to be seen in this area, here they did not constitute the kind of hazard that they had in the Chobe. There was an abundance of forage and water in the Moremi, so the elephants were usually to be seen placidly grazing in the forest. We seldom saw them traveling in herds within their protective phalanx of bulls. A rather different kind of traffic problem was caused by giraffes, which are very plentiful in the Moremi. Whenever we encountered them on the road, they would trot ahead of us with their characteristic swinging lope. It was a comical sight, following along behind them, with their massive hams swaying back and forth high above the windscreen of the Land Rover, but you could follow them for miles without them getting off the track. The only way to clear the way was to rush them a bit, tooting lightly at them with the horn and roaring the engine. Just when it seemed like the bumper would nip them on their heels, they would bolt off to the side. This maneuver involved a certain risk of collecting giraffe flops on the windscreen.

When we had first entered the reserve at Khwai, we had been warned to be on the lookout for rhinos. In the previous month some twenty black rhinos had been brought in from the Kruger National Park in South Africa, one of their last refuges, in an attempt to reestablish a herd in the Moremi. They had been kept in a pen for a few weeks to acclimatize them to their new habitat, and had only recently been released. We were told that they were probably all hiding in the deep bush and it was unlikely for us to see any of them, but that if we did, to be very careful, because they were likely to be very unstable and disoriented. Since none of us were keen to encounter a rhino even in a stable condition, we were kept in a high state of alert while we were driving on the tracks.

In a remote spot way out near the tip of the peninsula we came upon the pen where the rhinos had been kept during their initial internment. It was a large enclosure with walls constructed of earthen berms, ten feet thick and eight feet high. It contained big piles of partly consumed hay and was littered with dung. At one end there was a huge ramp where the beasts had been herded down from the trucks that had transported them from South Africa. At the other was an equally massive gate, standing wide open. The robustness of these structures vividly evoked a sense of the power of the animals that they had been meant to contain. The emptiness of the stockade cast a foreboding eeriness over the scene. You couldn't help but glance nervously around the surrounding forest.

As we drove away, I asked Thomas, "What do you do if you meet a rhino on the road?"

"Stop. Turn around."

Dumb question. I had once seen a picture of a car that had met a rhino on a jungle road and had been completely demolished. Fortunately, we never did run into one of those fearsome beasts, but the possibility kept us on our toes as we made our daily rounds of the instruments. Our encounters with wildlife always came from more unexpected quarters. One day Teddy and I returned from servicing an instrument to find the whole crew standing around the pickup, watching Robert pounding on the rear fender with a big stick. "What in the hell is going on?"

"Big snake in da wheel, boss."

Very cautiously, I peeked under the rear of the pickup. Sure enough, there was a huge python, about as wide around the middle as my leg, wrapped three or four turns around the axle and scrunched up against the wheel hub. Try as we might, by poking and prodding and pounding, it wouldn't budge. "What the hell," I said, "Let's go. It'll fall off somewhere." At our next stop, we looked again, and it was gone.

As I was getting dressed one fine morning, I shook my boots to get the sand out, and a scorpion fell out too. It marched around the tent floor with its tail and stinger up, threatening me pugnaciously. I dispatched it with a shovel and scooping it up, took it over to show to Deacon. It was the biggest scorpion I'd ever seen, big and fat and almost three inches long, body and tail.

Deacon looked at it with a start. "Dat one, he very bad, boss. He can kill de ox. You got to find de udda one."

"What other one?"

"Dey always live wid de mate. If you find one, der always anudda one nearby."

Teddy and I carefully searched the tent, poking at everything with long sticks. We finally found the scorpion's mate under a pile of dirty clothes in the corner. From that moment on our tent acquired the tidiness of a monk's cell, and we took to carefully searching our sleeping bags and cots before turning in at night.

On our long boat trips, we tried our luck fishing but soon found that our light spinning equipment just wasn't up to the sport to be found in the swamps. I made my first attempt on a beautiful lagoon where Simon and I had stopped to take our lunch. I took the boat out alone and within a few minutes managed to land a couple of good-size bream. Casting once again, there was a riffle and silver flash as something hit my lure like a torpedo and took off with it across the lagoon at an incredible speed. The line was screaming out of my reel, but before I could snub down on the drag, it went dead. Uh-oh, it was coming back at me! I started cranking in line as fast as I could. Before I could even make a dent in the slack the silver streak came shooting back under the boat and "pow," the line broke against the side of the boat. A little stunned, I rowed back to shore, where Simon was grinning. "It de tigerfish. You can't catch de tigerfish wid dat little ting."

The same afternoon, Simon demonstrated how it was done. His fishing equipment consisted of a three-foot piece of steel wire about as thick as a coathanger with a big treble hook on one end and the other attached to about fifty feet of clothesline. The lure consisted of a piece of beer can looped onto the end of the treble hook. He spun this rig around his head like a lasso and cast it out into the pool. On the third cast he got a tigerfish strike. Pow! He lost the works. He looked a bit sheepish retrieving the remains of his clothesline.

When I told Teddy this story it stimulated his sense of fishing ingenuity. He now considered it a personal challenge to catch a tigerfish. He spent hours in the equipment tent beefing up his fishing tackle, making up steel leaders and heavy-duty hooks out of odd bits of gear that he found in his repair kit. In spite of all his efforts, he succeeded only in losing a lot of tackle. Tigerfish were just way out of our league.

The Khwai River Lodge was in a beautiful setting, in a shady grove fronting a big lily pad–covered lagoon. On one side it had a grass landing strip where the air safari planes could come in. The hotel itself was what the architect must have called "rondavel-inspired." It consisted of several large, one-story, circular buildings with plain whitewashed walls and conical thatched roofs. We drove by it every time we transited from the north to the south end of the Moremi, and once in a while we would stop in and get a cold soda or sandwich at the bar. The bar was an outdoor lanai attached to the dining room that had an open front looking out onto the lagoon. It was a magnificent view, with a herd of lechwe grazing along the bank and lily-trotters striding across the lily pads with their big floppy feet. The first time we went there the black barman refused to serve the crew. I got into an argument with him about it, which was settled when the white manager arrived and instructed him to go ahead and serve us. He then took me aside and told me that it was OK this time but that it wouldn't be a good idea for us to bring our crew in when the lodge had guests. They seemed to have very mixed-up racial rules in this country. There seemed to be different rules in places frequented by well-heeled white tourists.

We did go there once when a tour group was in residence. We didn't realize it until we rounded the building and found the dining room filled with people having lunch. It was really a first-class affair. The linen-draped tables were set up with full services of silver and crystal complete with centerpieces bursting with flowers. Coveys of waiters rushed about serving the guests, who were dressed in the sort of clothes you might expect to see at an upscale Maui beach resort. We could tell from their conversation that it was a group of wealthy Americans on an air safari tour. The sight seemed so incongruous to our eyes that we couldn't help but gawk a bit, but they really gawked back at us. As Teddy, Blackie, and I waited outside the bar for our case of Pepsis, several of them actually came up and brazenly took our pictures, as though we were some kind of picturesque tableau brought in by the lodge for their lunchtime entertainment. I could just imagine them back in Wichita or someplace, showing our pictures to friends they had invited over for a slideshow of their great African trip. "Our guide told us that these people were some kind of scientists from New York, out there studying earthquakes. Can you believe it? They just looked like a couple of old desert rats to me. Didn't they, Marvin?"

Driving into Maun one day for supplies, we came across an unusual sight on the main road south of Khwai—a Volkswagen Beetle stuck in the sand. Even in Maun one scarcely saw any vehicles that were not in the four-wheel drive class, and certainly not traveling alone this far out in the bush. We pulled up and asked the driver if he needed any help.

"Ja, ja. Just stuck in a bit of sand, hey? Could you have your boys give me a push?"

We all piled out and started to push the little car, which was pretty mired down. Just as we got it moving, he stuck his head out of the window and said, "Wait, wait! You cawnt be doin' that, man."

"Huh? Doing what?"

"You cawnt be doin' the pushing, man. Just let your boys take care of that."

Shit! "Just steer, OK? We'll take care of the pushing."

After we freed his car, we went up to talk to him. I was surprised to see him climb out of that little car. He was a huge Afrikaner, built like an ox. "You need any more help?"

"Thanks, you blokes. She's right, now. Can you tell me how far it is to Kudomane? I need to get petrol there."

"Kudomane? Where the hell is Kudomane, Ted?"

"Beats me. Never heard of it."

"Sorry, we've never heard of Kudomane. Are you sure you're on the right road?"

"Ja, ja, no problem about that. Here, I'll show you on the map." He pulled out a south African road atlas and showed us the map of northern Botswana. The road we were on, up through the Chobe to Kasane, was indicated with a thick red line, as though it were a major highway. Sure enough, just south of the park entrance there was a black circle on the route, labeled "Kudomane." I noticed that further along was another black circle, as though indicating some prominent town on the road. It was labeled "Ngwezumba."

"There's nothing at Kudomane. It's just a locality. There's no town there. Maybe there are a couple of huts, but there is certainly no place to get any petrol. You'd best go up to Khwai, along here, see. There's no petrol station there either, but you can probably talk them out of some petrol at the Lodge. Where are you headed, by the way?"

"Kazungula. Goin' to meet up with me mates there. They're coming over from Rhodesia."

I was flabbergasted. "You can't get to Kazungula from here in this car. You got to cross the whole of the Chobe. Even if they let you through, the road up there gets much worse than the one here, where you just got stuck. There are no services on that road: no petrol, no food, no water, no people, no nothing. Just elephants. Even if you managed to avoid any breakdowns, you would never make it before dark. And you can't be out in the Chobe after dark."

"Naw, naw. I checked with the Automobile Association in Pretoria, see. They made this map, see, and they know this area. They said it was no problem." And with that, he got in his car and drove off.

Ted and I just looked at each other incredulously. "I think we just met a Van der Merwe, Ted." Van der Merwe is the proverbial block-headed Afrikaner. We had been told countless Van der Merwe jokes since we had been in Botswana. Even the Afrikaners tell them. I heard the rest of the story a few weeks later at Riley's. This Van der Merwe's Volkswagen broke down halfway between Savuti and Ngwezumba, stuck in the sand with a ruptured oil pan. Within three hours the Savuti water truck had come along on its monthly rounds and towed him the rest of the way into Kasane. He certainly had had the Afrikaner's angel sitting on his shoulder.

While we were still in Ngwezumba Teddy had received a telegram from Judy, his wife, telling him that his Navy Reserve group had been complaining because he had missed so many weekend meetings and that he had better show up soon or risk getting kicked out of the Reserve. He took this warning seriously, so I arranged for him to leave a few weeks early and for Dave Hutchins to come up from Lobatse to take his place. I kidded him that actually he was just trying to escape the clutches of his Groovy Queen, which he vehemently denied, of course. Actually, I wouldn't have minded going with him. We had been out for more than three months and were both getting a bit field-weary. I had to stay for an extra couple of weeks, though, to wrap up the work in Moremi and to train Dave Hutchins for a last little leg of the project that I had just worked out. I had really wanted to operate a network out on the western edge of the delta, in the hopes of picking up some activity from the faults on that side. What with all of our other problems, Teddy and I had run out of time, but Dave had volunteered to do it for us. I had managed to coerce my colleagues back at Lamont to let me keep the instruments a while longer than scheduled (on the "possession is nine-tenths of the law" principle), so we planned to go ahead and do that as long as the rains continued to hold off.

Dave drove up from Lobatse and we met him at Riley's, where we had a quiet little farewell party on the eve of Teddy's departure. The next morning we took him to the airport to catch his charter flight to Gaborone. While we were waiting for Don to warm up the Beechcraft, I noticed a battered old DC-3 parked on the far side of the dusty apron. There was a line of African men in tattered clothing waiting to board, each with his small cloth parcel of belongings.

Joe stood at the head of the line, checking people off on a clipboard. He was shipping off a consignment of labor for the mines in South Africa.

I've been in those mines. The gold mines of the Witwatersrand are the deepest in the world: The working faces are at a depth of nearly two miles. In the one I visited, you go underground by means of a skip, which is like an armor-plated roller coaster train that descends on rails at an angle of 45 degrees. They let you go free-fall for the first half mile, which gives you an appreciation of what gravitational acceleration is all about, and then start hydraulically braking, storing the energy to lift the train on the adjacent rails. The working faces are at the end of stopes which are slots cut into the rock at a slope of about twenty degrees. These are fifty to sixty inches high at the working face, just slightly wider than the gold-bearing reef. The face is continually being advanced by an endless succession of drilling, blasting, and mucking away of the ore and waste. The rock in the Witwatersrand is extremely strong and brittle, which is what allows them to mine so deep, but even so the roof is continually cracking, so that a hundred feet back from the working face the stope is scarcely ten inches high, in spite of wooden and steel pillars installed to support it. The mine is continually closing in on itself, so they literally have to keep mining to keep it open. It also has to be continuously air-conditioned: The natural temperature at that depth is about 160° F, about that of the uppermost shelf of a sauna. It is truly a hellish working environment. While gazing at the human cargo boarding the DC-3, I wondered if any of those men, those sons of the open veld, had any idea what they were heading for.

After seeing Teddy off, Dave and I returned to the Moremi camp. I had just begun to get my notes together, and Dave had brought up a collection of books and papers I had requested from the Geological Survey's library in Lobatse. I spent the next few days collecting my thoughts and observations and checking them with other information I found in those publications. At this point I was no longer concerned with the immediate goals of our project. I had already done enough preliminary analyses of the data we had collected to see that I would be able to provide a useful report to the F.A.O. on the seismic hazards in the Okavango, and that we had enough data to prove that the Okavango was indeed at the tip of a nascent branch of the African Rift System. I was now more concerned with setting down in my notes some thoughts concerning the broader questions of the mechanism of continental rifting and the forces that drive it.

There is no doubt that hotspots can play a major role in the initiation of continental breakup. There is plenty of evidence in Africa for that. In the most recent example, the Afar hotspot was associated with the rifting off of Arabia and with the initiation of the East African Rift System, but there are several earlier examples as well. In southern Africa there was a series of intracontinental basins that for more than 100 million years from the Silurian through the Triassic periods collected a thick sequence of sediments called the Karoo System. During the Jurassic these were uplifted to form high plateaus, which were partially rifted and finally filled with a massive outpouring of flood basalts. I had seen a little piece of this in the Ngami Basin. These later events are thought to mark the arrival of one or more hotspots, huge, partially molten plumes rising from the deepest part of the mantle, perhaps as deep as the core-mantle boundary. The plumes are thought to underplate the continent with magma that solidifies and, being less dense than the surrounding mantle, causes it to gravitationally dome upward. As the doming becomes more intense, stretching the crust, rifting occurs, and then the basaltic magmas erupt to fill the rifts, spilling over in places to blanket the surface of the plateau.

The arrival of these Karoo hotspots immediately preceded, and may have caused, the breakup of East and West Gondwanaland, the southern half of the supercontinent of Pangaea, with the splitting off from Africa of India, Madagascar, and Antarctica, and the formation of the Indian Ocean. Karoo-age flood basalts are found both in northern Madagascar and in Queen Maud Land in the adjacent part of Antarctica. Volcanic ridges on the seafloor between South Africa and Madagascar and on to Antarctica may mark the track of one of these hotspots, which may currently lie, in a much depleted state, beneath Marion Island between South Africa and Antarctica. The other appears to be emitting its last trickles of magma beneath Bouvet Island in the South Atlantic.

Another hotspot, which arrived farther to the west some 50 million years later in the late Cretaceous, can be much more clearly shown to be related to the splitting off of South America from Africa and the formation of the South Atlantic Ocean at that time. That hotspot, currently located beneath Tristan da Cunha, has left clearly recognizable tracks on the seafloor that symmetrically trace it back to both continents. To the east, the Walvis Ridge connects it to Namibia where flood basalts of late Cretaceous age are found in the Etendeka district; to the east the Rio Grande Rise traces its path back to the Paraná basin in Brazil, where there is an enormous field of flood basalts of the same age.

Rifting itself, though, clearly doesn't require the forces associated with a hotspot. The rift in the Okavango, as well as many of the other East African rifts, is simply too far from the Afar to be affected by it. Certainly no doming preceded the Okavango rifting. Although the Kalahari is high, at an average elevation of just over three thousand feet, this is a relic of the Karoo uplift, not of any recent doming that might be associated with the present rifting.

It seemed as though the initiation of continental breakup required the extra energy of a hotspot, but once it was initiated, rifting could proceed under the action of lesser and more distant forces. I was reminded of splitting logs for firewood. If you have a mature log of a hardwood like oak or maple, two or three feet across, you first have to saw it into two-foot lengths. You'll find then that you can't begin splitting it with an ax unless you are content with hacking away at the outer edge. The trick is to start with a wedge, which you insert about half to two-thirds of the way out from the center and drive in with a sledgehammer on a line tangent with the rings. Once you make that first split, it releases the hoop stresses that bind the wood together, and the rest can be split easily with an ax. The hotspot is perhaps analogous to the wedge. Once the breakup of the continent is initiated, rifting can propagate across the continent more readily and at great distances from the hotspot. The ax cannot split through burlwood, of course; one must split around the knots in the wood. Similarly, the rifts cannot propagate readily through the cratons. They follow the old tectonic grain of the continent, along the greenstone belts and old rifts like the Karoo rift in the Ngami Basin.

I am not sure why the initial breakup of a continent requires greater forces than does the subsequent rifting. I have some ideas, but they are not yet developed fully enough to be set down on paper. If I ever do figure this part out, I will have solved the whole problem. Then I will write a paper on it, and be done with it. As it is, it is still on the list of unsolved problems, and is much debated today, with little sign of a resolution in sight.

While I was thus ruminating, Dave and I continued carrying out our daily chores. Camplife without Teddy seemed empty and listless. After Dave had been in camp for a few days, Deacon started to show signs of reverting back to his old British cooking habits. Dave went rooting about in the tucker box and was delighted to see that all the tinned treats that he had bought in Mafeking were still there. These were all the things that Teddy and I had been shuffling aside countless times in the past three months. Dave came out of the tent triumphantly holding up a pie-shaped tin. "Steak and kidney pie. Great, you saved it for me! Let's have it for dinner tonight."

I remembered this tin. On several occasions, desperate for some gustatory novelty, I had suggested it to Ted. "Eeyew, kidneys. Never touch the things. They taste like pee."

Dave gave the tinned pie to Deacon to heat up for our dinner, and we sat down to play some cards. Some while later we were startled by an enormous BOOM! I jumped up, and looking behind me, saw Deacon running around in front of his fire holding a towel over his head, while above him the cast-iron lid of his kettle languidly spun in a parabolic arc beneath a tree that was now festooned with shreds of steak and kidney pie.

"Damn," said Dave, "I forgot to tell him to puncture the tin."

Life with Dave settled down to a dull routine. He had brought his cribbage board along, and taught me how to play the game. It is a fairly simple game, easily mastered. We played for a penny a point "just to make it interesting" and although I fell massively into debt in the beginning, I caught on to the simple strategy soon enough and ended up owing only something like 25¢ after two weeks of steady play.

The evening's card playing was enlivened with running commentaries by Dave about the joys of British pub life. His great goal in life was to buy a pub somewhere in a small rural village and to settle down as the village publican, which to him was the most hallowed of all occupations. This geology stuff was just something he was doing in the meantime. His plan in coming to Botswana was that he would be able to live cheaply and save most of his salary, so that by the time his contract was up he would be able to return to Britain with enough for a down payment on a small pub. He had it all planned out in great detail, although he admitted that what with all his trips to Gabs and down to Jo'burg he hadn't been saving at a rate quite up to his expectations. In fact, he had saved hardly anything, which is why he was happy to come to the field rather than spend Christmas in Natal with a bunch of his mates.

He regaled me with all sorts of arcana about the merits of various ales, beers, ciders, and pub meals. He had intimate knowledge of the rules and mode of play of an amazing variety of pub games, some of which haven't been manufactured since Victorian times. He had his whole pub set up in his mind's eye: which games were where, which drinks were on draught and which were in bottles behind the bar, what was on his permanent meals menu and what were his daily specials. His was an independent pub, of course, bravely withstanding the dictates of the powerful breweries and selling only good live beers made by honest craftsmen in the old way.

We ate his tinned Christmas dinner prematurely, since Dave decided it wouldn't do to eat it alone. It was a cooked stuffed goose, shaped like the tin it had been squeezed into, which when heated up and served with roast potatoes wasn't all that bad. The plum pudding with hard and rum sauces was excellent. While we were eating it, he described a typical Christmas at home. "After we open the presents in the morning, we all go round to the pub: me da, me brothers, and me, while the women cook the meal. At noon, we come home for the meal. It's like this, only better. Not in a tin, or anything. Me ma is a great cook. After the meal we have a bit of a kip, and then it's off to the pub again for the evening."

"Sounds like a very spiritual Christmas," I remarked.

Two weeks of this passed quickly and uneventfully, and my departure date arrived anticlimactically. We had a little celebration at Riley's, in which I bought the crew a beer, except Deacon, who took a Groovy. There were no speeches. I just went around and shook everyone's hand, and thanked them individually. I had developed a great respect for them. Each of them, over the past three months, had carried out his duties skillfully and without complaint, even under the sometimes harrowing circumstances that we had experienced. They were all, no matter how modest their skills, professional fieldmen, who had done their work as part of the team. Although once or twice they had challenged my leadership, I had always felt that they had been only testing the water. They would follow me as long as I could maintain their faith in my own professionalism. Blackie stood at the end of the line. I took his hand and thanked him with great sincerity of feeling. He said, in return, "Thank you, boss. You good boss for us. If you come back to Botswana, we want to work for you again." I was touched.

I had a final pep talk with Dave. He was to take the instruments out to the west of the Okavango and run them there until the New Year before shipping them back to New York. I urged him to put one in the Tsodilo Hills, which was out in that area, but I didn't really think he would actually do it. It was a Bushman hill, and I didn't think he would be able to get the crew to go there, even if he wanted to himself. Besides, I could tell that Dave didn't really have his whole heart in the job. Half the time he was wandering around daydreaming about his imaginary pub.

The next morning we went out to the airport so I could catch my charter flight to Gaborone and on to New York. Joe was there once again with the DC-3 and another consignment of contract labor for the mines. While I was waiting for Don to wind up the Beechcraft, I noticed standing next to me a man waiting to board the plane for the mines. He was tall and slim and very black, and was wearing a threadbare pair of khaki shorts and shirt, and sneakers held together with duct-tape. In one hand he held a small tattered cardboard suitcase, and in the other a Bushman's harp, a block of wood with some nails driven through, bent over and flattened out to form a crude plucking scale. He was plucking out one of the strange, endless, atonal Bushman rhythms, while staring sightlessly out at the middle distance, as though his soul had already half fled his body. I shuddered inwardly, thinking about what different destinies our flights would be taking us to. I turned, and hoisting my duffel and grabbing my satchel of data, walked out to my plane.

≈≈ *Epilogue* ≈≈≈≈≈≈≈≈≈≈

Back in the comfort of Lamont, after a month or so of rest and recu-
peration I sat down to analyze the data. Our haul had been meager
enough, but it turned out to be just sufficient to nail down the main
elements of my hypothesis. We had recorded microearthquakes on
enough instruments to get good locations for about fifty of them.
These showed a band of seismic activity running northeasterly from
just south of Lake Ngami all the way to the Zambezi, following the
activity stepping over to the east to join up with the Kariba Gorge
activity. From these data I was able to construct two composite focal
mechanism solutions, which showed that these earthquakes were
occurring on northeasterly striking normal faults: faults with
mainly vertical motion, with the northwestern side moving down
with respect to the other. This is precisely the type of motion that
would be expected if these were bounding faults of a nascent rift
valley, damming up the Okavango.

Combining these results with data obtained by the Rhodesia Seismic Network, it was easy to show that this was the terminus of an
arm of the rift extending from the Lwangwa down to the Kalahari.
Here was evidence that a rift could propagate across undeformed
country without a trace of prior arching or volcanism. I published
these findings in a short paper in the *Geophysical Journal of the Royal
Astronomical Society* in 1976, with Ted and Dave as coauthors. The
same results were presented to the F.A.O. as a report the year earlier.
For their report I also included a seismic hazard analysis, in which I
calculated, from the microearthquake activity level, the statistically
likely rate of recurrence of magnitude 6 earthquakes such as those
which had occurred in 1954, and which would be likely to affect the
hydrological regime of the Okavango Delta. I included an analysis of
the expected effects on drainage patterns. So each of my tasks, the
scientific one and the consulting one, were completed satisfactorily.

In the ensuing twenty years, there has been little or no progress on the main problem: the cause and mechanism of continental rifting. Although this is still a big problem in geology, it is a stagnant one. In the absence of a breakthrough, it is a problem that has been tabled, with people working just on its fringes. In that atmosphere, our little paper was relegated to one of regional interest only: it was not published at a time of great controversy over rifting, when it might have attracted greater attention. Its citation record reflects that; it gets referenced every couple of years by someone working on the tectonics of southern Africa. It is basically a paper that reports a few facts in a little-studied part of the world, and hence will be infrequently cited for as long as its facts do not get replaced by new and improved facts. But it also established a few larger facts: the most important being that rifts can propagate without the driving force of a mantle plume. The velocity of the rift propagation which we could now estimate is also quite interesting. If rifting initiated in the Afar 24–30 million years ago, and has just reached the Okavango, some four thousand kilometers to the south, then even if one assumes a constant propagation rate, it must have been lengthening at something like ten centimeters per year, a rate comparable to plate motion velocities. If, as is more likely, it has slowed down with time, it must have started out growing much more rapidly. These are numbers which are very hard to come by, and which will eventually be critical in testing models of continental rifting.

If and when a breakthrough on continental rifting is made and it becomes a hot topic again, our little paper may enjoy a brief day in the sun. As it stands I consider the Botswana project a minor sidelight in my career. It was not one of my major accomplishments, but neither was it a failure. I have spent much more time than that on other projects with much less return. But the Botswana fieldwork was about much more than science or career. Some things in life are worth doing solely for the experience.

The past twenty years have been prosperous ones for Botswana as a result of its political stability and the consequent development of its mineral resources, which have proven to be just as rich as expected. Orapa is indeed now the world's largest diamond mine. The country now boasts a paved network of principal roads, and the towns along the southeastern corridor bordering the old Rhodesian Railroad contain the modern amenities one might expect of a rapidly developing country. This great mineral wealth masks, however, underlying problems more typical of other parts of Africa. During this same twenty years the human population of the country tripled, so that in many areas the land is suffering from the pressure to supply ever larger amounts of forage and firewood. Botswana is no longer a net exporter of food.

The Okavango Delta and the Chobe National Park have been spared this sort of exploitation; they have instead been maintained as reserves and foster a flourishing tourism industry. There are now many more opportunities for the tourist to see these areas than there were when Teddy and I were coping with the primitive conditions I have described. The "highway that Kissinger made" has completely opened up the Chobe by road. It is now possible to take motor safaris all the way across the park from Kasane to Maun. A dam has been built at Ngwezumba, producing some pools, more reliable than the old water hole, which support great wildlife herds and are a major tourist attraction. There is even an occasionally used safari camp at Ngwezumba, not far from the camp we so precariously occupied for a month! In spite of a great drought in the '80s, the wildlife populations have rebounded and, if anything, are stronger and more diverse today. The Chobe elephant herd, free from the predation of ivory poaching that has decimated their species in other parts of Africa, is now over fifty thousand strong, and the government faces the hard decision of whether or not to cull the herd to prevent its damaging the delicate veld habitat by overgrazing. Meanwhile, in the Okavango, the crocodile population is now on the way to a full recovery.

Teddy and I still enjoy a partnership in science. At the time of this writing he is in Siberia, somewhere to the north of Irkutsk, leading a field team as part of a joint U.S.-Russia project to make precise geodetic measurements of Russia with GPS technology. These data will eventually lead to better measurements of the deformation of Asia resulting from its collision with the Indian subcontinent along the Himalayan suture zone.

Christopher Scholz is Professor of Geological Sciences at the Lamont-Doherty Earth Observatory at Columbia University. He is the author of *The Mechanics of Earthquakes and Faulting,* and many technical articles which have appeared in leading science journals including *Science* and *Nature.*